# Communication in Construction Teams

# Also available from Taylor & Francis

## Communication in Construction: Theory and Practice

A. Dainty, D. Moore and M. Murray

Hb: 0-415-32722-9
Pb: 0-415-32723-7

## Risk Management in Projects, 2nd edition

M. Loosemore, J. Raftery, C. Reilly and D. Higgon

Hb: 0-415-26055-8
Pb: 0-415-26056-6

## Project Planning, and Control

D. G. Carmichael

Hb: 0-415-34726-2

## Construction Project Management

P. Fewings

Hb: 0-415-35905-8
Pb: 0-415-35906-6

## HR Management in Construction Projects: Strategic and Operational Approaches

M. Loosemore, A. Dainty and H. Lingard

Hb: 0-415-26163-5
Pb: 0-415-26164-3

Information and ordering details

For price availability and ordering visit our website
**www.tandf.co.uk/builtenvironment**
Alternatively our books are available from all good bookshops.

# Communication in Construction Teams

**Stephen Emmitt**
and
**Christopher A. Gorse**

Routledge
Taylor & Francis Group
LONDON AND NEW YORK

First published 2007
by Taylor & Francis

Published 2015 by Routledge
2 Park Square, Milton Park, Abingdon, Oxfordshire OX14 4RN
711 Third Avenue, New York, NY 10017, USA

First issued in paperback 2015

*Routledge is an imprint of the Taylor and Francis Group,
an informa business*

Typeset in Times by
Integra Software Services Pvt. Ltd, Pondicherry, India

*British Library Cataloguing in Publication Data*
A catalogue record for this book is available from the British Library

*Library of Congress Cataloging in Publication Data*
Emmitt, Stephen.
Communication in construction teams / Stephen Emmitt and Christopher
Gorse.

p. cm.

Includes bibliographical references.
(hardback : alk. paper) 1. Communication in the building trades.
2. Construction industry--Management. I. Gorse, Christopher A. II. Title.

TH215.E447 2006

690.068'4--dc22                                   2006005623

ISBN 13: 978-1-138-97118-9 (pbk)
ISBN 13: 978-0-415-36619-9 (hbk)

*To Ruth and Pam*
*None of this would have been possible without your help and support.*

# Contents

## 9  Discussion of the findings                                      224

## 10  Conclusions and recommendations                               240

# Preface

The importance of effective communication within construction teams is well known, yet published data regarding the manner in which individuals interact within the temporary project team is scarce. Thus, managers and researchers have little other than anecdotal evidence on which to develop effective tools and methods to facilitate construction communication and hence better the performance of the project team. Our aim in writing this book is to provide an insight into communication in construction teams based on extensive research from live construction projects.

In this book we provide a comprehensive overview of the literature on interpersonal communication and a critical review of various research methods previously used both within and outside the construction management field. The review leads to a focus on Bales' interaction process analysis (IPA), a proven and robust research tool, which has been used successfully in many fields to collect interaction data. The IPA technique was successfully piloted before being used on a large-scale research programme that investigated the link between successful projects and the effectiveness of communication. Bales' IPA was used to categorise task-based and socio-emotional communication acts in a series of site-based construction progress meetings. This data was then compared with specific project performance criteria to see how communication affected the success of the project. The findings indicate that participants exhibit regular patterns of interaction, but there are different patterns of interaction associated with successful and unsuccessful projects – this is essential information for researchers and managers. Strengths and weaknesses of the research method are also discussed as an aid to other researchers in the field. Based on the empirical work a number of practical suggestions are put forward to assist all actors involved in construction projects, regardless of their position within the construction team. Our focus in this research is on site-based progress meetings, which are common to the vast majority of construction projects, and to which the main actors attend and interact during the course of the construction project. The data was collected in the UK, although our experience of other cultures tends to suggest that our findings are applicable to those working elsewhere.

The book has had a long gestation period. As practitioners, we were independently dealing with aspects of communication through our daily management activities, trying to implement procedures to improve the effectiveness of communication within construction projects. In 1996, we both moved into an academic environment and started working together on aspects of construction communication, starting with small research projects, a completed doctoral thesis (of Christopher A. Gorse) and continued data collection that has led to this book. We have systematically and deliberately integrated our research findings into our teaching since 1996 to enrich our work in the softer side of project management. Our earlier book *Construction Communication* (2003, Blackwell Publishing) sets out some of the fundamental communication issues common to the successful management of construction projects. The data published here is complimentary to our earlier book and we would urge the readers to consult both to ensure a rich understanding of the issues facing all actors in construction, regardless of their position in the project team. Consistent with the style of our earlier book, we have attempted to write in a simple and informative style. To the best of our knowledge this is the only book on communication in construction teams that is based on research collected from meetings during live projects. We hope it helps to shed a little more light on an area too often taken for granted.

We are extremely grateful for the help and encouragement we have received over the past decade while researching various aspects of construction team communication. We are particularly indebted to the organisations and individuals that gave us permission to enter their commercially sensitive world to collect data and ask impertinent questions. As promised, you will remain anonymous – but thank you all the same. We would also like to extend our gratitude to our Masters students at the Technical University of Denmark and Leeds Metropolitan University who have been keen to engage in discussion about the softer side of project management. The Johan Hoffmann og Hustrus Mindefond provided the funding for Stephen Emmitt's professorial chair at the Technical University of Denmark and hence the space that helped us to realise this complex undertaking. Finally we would like to thank the anonymous individual who suggested to Tony Moore at Taylor and Francis that our research findings should be published as part of a research series: here it is.

Stephen Emmitt
Christopher A. Gorse

# 1 The social life of construction projects

Communication in construction teams, or rather the effectiveness of communication in construction teams, is a significant factor in the successful completion of construction projects. Communication between the organisations and individuals temporarily brought together to realise a constructed work is undertaken constantly but remains an aspect of construction that has, to date, received very little attention from researchers. A number of government-led reports have consistently drawn attention to the difficulties caused by the organisational systems in which construction teams operate. The inadequacies reported seem to stem from poor interaction practices. The nature of interaction affects the strength of relationships between the actors and ultimately colours their ability to transfer knowledge and appropriate task-based information to complete projects successfully. Team building, the discussion and subsequent sharing of values, resolution of minor differences and conflicts, asking questions and the creation of trust between construction team members are just a few of the factors that are crucial to the smooth running of projects and which are reliant on the ability of the actors to communicate effectively and efficiently. It follows that the interaction of multidisciplinary parties within the construction team should be of real interest to researchers and managers alike. Despite this, there appears to be very little research into, or indeed knowledge about, what actually happens in practice. Much is spoken and written about the importance of project communication; unfortunately much of this is largely based on anecdotal evidence rather than on the findings of applied research in the workplace.

The first specific review of communication in the UK construction sector appears to have been undertaken by the Tavistock Institute with two publications, *Communications in the Building Industry: The Report of a Pilot Study* (Higgin and Jessop 1965) and *Interdependence and Uncertainty* (Building Industry Communications 1966). These two publications tentatively follow some of the issues raised earlier in Simon (1944), Phillips (1950) and Emmerson (1962) reports. It is also possible to see some parallels with the work of the economist Marion Bowley (1960; 1966), who was concerned with innovation in the UK construction sector. The Tavistock publications attributed many of the problems to the separation of the design and production teams, a theme that still persists in the research literature. More recent reports initiated by the UK government, for example Latham (1993, 1994) and Egan (1998, 2002), follow a similar pattern in urging improvements and suggesting ways in which the construction sector should be organised. Similar trends can be seen in many other countries, for example Denmark (Kristiansen *et al.* 2005). The construction reports do not address aspects of communication explicitly, but in the message to improve existing practices it is implicit. In a comprehensive review of the communication literature, Emmitt and Gorse (2003) found a paucity of scientific research. With the exception of a small number of doctoral studies the published research was based on questionnaire surveys and interviews, thus restricting the data gathered to actors' perception.

Since publication of the early work there has been a rapid development of information technologies (ITs) and information communication technologies (ICTs). Much attention has been paid to the impact that the new technology has and will have on our working lives. In architecture the increasing sophistication of technologies and computer aided manufacturing has allowed for more exciting designs and in some cases a redefinition of how buildings are procured. Current emphasis is firmly on digital information and improved communications through project web technologies, which provide a tool for better exchange of information between project participants. However, doctoral research has shown that there is no correlation between the use of project web tools and improved communication between team members (den Otter 2005). The message in den Otter's research is that we need to be careful when implementing ICTs and remember that the use of new technologies does not necessarily ensure better performance. Technological advances in communications are very welcome; however, the lure of technology has detracted from the softer, and arguably more complex, issue of interpersonal communication in construction teams. It is the interpersonal, often informal, communication that forms the glue between individuals and organisations contributing to projects. Few researchers have attempted to observe or examine the nature of construction communication as it happens in live projects (see Chapter 2).

Another factor worthy of consideration is the manner in which we organise design and construction projects. Government-led reports have a tendency to promote a particular approach, for example partnering, supply chain management, lean thinking etc. This (temporarily) skews research into specific areas identified in the reports and some of the more complex underlying issues, common to all construction projects, tend to be overlooked. There is, of course, no universal 'answer' to the challenges facing construction participants, instead there are a number of underlying factors common to each project, regardless of context. The dominant factor is the temporary interaction of people and our collective ability to use communication media effectively in dynamic situations. It is the social life of construction projects and the formal and informal relationships that help and sometimes hinder the delivery of a successful project. Emphasis for researchers and practitioners alike should be on the 'contact' between actors rather than the legal contract. Researching and hence developing a better understanding of how participants interact during construction projects and how they communicate within the context of the construction team would, in our view, be a significant step forward. To start understanding this complex area, it is essential that research is rigorous, based on real observations and/or can be applied to real contexts. Most importantly the research should be published in the public domain.

## Communication

Communication is arguably *the* one aspect of project management that pervades all others, for without effective communication between the participants the project team cannot succeed in realising its objectives. Key management competences of leadership and decision-making are founded on good communication skills. Construction projects are undertaken in a dynamic social system, nothing is particularly stable for very long, and uncertainty and interdependence are constant factors, regardless of size or complexity. It is the dynamic interaction of the various actors to a construction project that constitutes a process of communication.

We use the word 'communication' throughout this book in the way it is used in the majority of the communication literature, i.e. *communication* is the sharing of meaning to reach a mutual understanding and to gain a response – this involves some form of interaction between the sender and receiver of the message. It is important to note that we do not use the term 'communication' in the way it is in some of the construction project management literature, where to 'communicate' is simply to transfer information to another party. Here, the opportunity to ask questions to reach mutual understanding is rarely available (and we wonder why we get into difficulties). Indeed, with some notable exceptions (e.g. Moore, 2002) much of the construction project management literature fundamentally fails to deal with the interaction of disparate actors and the manner in which they communicate in construction teams.

The creation of meaning between two or more people at its most basic level is an intention to have one's informative intention recognised (Sperber and Wilson 1986). Informing someone by any action that information is to be disclosed is considered to be an act of communication. There are many ways that people make their introduction and let others know information is about to be disclosed, or by their non-verbal actions set a context for discussion. A client's representative could, by way of a raised eyebrow or nod in the direction of others, send a non-verbal signal that contains significant information to alert a construction manager to act; for example, to stop site workers from engaging in inappropriate behaviour. The message and clues can be very subtle, but still convey considerable information to the receiver of the message. Status is another factor. The mere presence of a person with reputation in a particular context or whose profession is known can send sufficient information to identify the nature of the discussion that is likely to take place. Specific examination of how professionals convey 'informative intention', how they capture the floor in a meeting and how they hold and maintain conversations encouraging others to hang on to the next word is useful and important if one wants to develop such skills. This book investigates task-based demands and emotions that are used to capture the interest or attention of others, but clearly there are other aspects of non-verbal communication that should be investigated.

Sperber and Wilson (1986) suggest that when people communicate they intend to alter the cognitive environment of the persons whom they are addressing. As a result, it is expected that the receiver's thought process will be altered. Communication performs much more complex tasks than simply letting someone know that we are about to send information. For understanding to take place most theorists claim that a background of shared social reality needs to exist. To engage in a meaningful communication we need to build on information and develop a context supported by cues and clues. These guide us to use subsets of knowledge and help us to link information together. Clues used come in many different guises. For example, Brownell *et al.* (1997) suggest that the appropriate mode of referring to something or someone in conversation depends on what common ground a speaker and addressee share. It is this common ground and the development of a shared understanding that makes communication possible. Similarly, a lack of common areas of understanding can create difficulties in communication.

Where there is a lack of congruent understanding, the speaker has to provide an infrastructure of contextual information onto which the new information can be built, developed and hence understood. During interaction the speaker has to access and build a framework of the other's knowledge and influence the way the recipient draws together the subsets of knowledge to make communication work. Using clues sent by the receiver of the message the sender may make assumptions of the knowledge the receiver possesses. This initial and often tentative interaction can be used to check whether early assumptions are correct at a critical stage in the development of a relationship. In practice, offering information, opinions, beliefs and asking questions help to develop an understanding of the other person's knowledge and beliefs, as does examining and analysing the verbal and non-verbal information presented (Bales 1950, 1970). Failure to exchange information and ask questions will decrease a person's ability to use the other's knowledge and thus considerably hinders effective communication (Brownell *et al.* 1997). The importance of cues and clues that are embedded in the context, body language and emotion play a key part in the development of understanding and relevance of exchanges. An important element of our research is to identify the different types of communication exchanges used by construction team members to access and present information, since this influences behaviour and affects the successful performance of tasks.

Improvements in the effectiveness of communication between individuals and groups can help to increase the performance of projects and organisations (Austin *et al.* 1993; Hill 1995; Azam *et al.* 1998; Shirazi and Hampson 1998; Swaffield 1998; Mutti and Hughes 2001; Swan *et al.* 2001). In a survey of 500 engineering firms by Gushgari *et al.* (1997), communication and listening were the first and second (respectively) on the project managers' list of most needed skills. A questionnaire survey, obtained directly from the workforce, found poor communication to be one of the problems experienced in the construction sector that results in serious production difficulties (Zakeri *et al*. 1996). Best practice in construction teams is characterised by open exchanges of information and good communication (Cook and Pasquire 2001), enabling responsibilities and ownership of tasks to be negotiated and shared (Macmillan 2001). Furthermore, where inadequacies in the communication process are reported, problems with construction activities arise (Building Research Establishment 1987; Cook and Pasquire 2001). Research undertaken by Taywood Engineering found that poor communication, lack of consultation and inadequate feedback are among the primary causes of construction defects (Betney 1997). Ineffective communication is a challenge not only for people involved in construction. In a report into the effectiveness of UK government procurement (the Kelly Report, 2003), similar concerns were voiced about the lack of communication within and between government departments and their suppliers.

## The construction team

In the context of this research it is necessary to be clear what is meant by terms such as 'team' and 'group'. Construction 'teams' are a loose grouping of interested parties brought together for a specific construction project. The team is composed of specialists operating in a disaggregated sector, each carrying different values and intentions to other team members. The reality is a series of individuals and groups working towards individual and group goals in a temporary social system that we term a 'project team'. (For a more detailed argument, see Emmitt and Gorse, 2003). Although the term 'construction industry' is used, the reality is that construction draws from many different disciplines, with many organisations also operating in other sectors; thus construction is not a particularly homogenous sector. Similarly, organisations are not as homogeneous or stable as we would like to assume. Contractors tend to rely very heavily on sub-contracted and sub-subcontracted labour. So do architects and engineers, outsourcing non-core activities to other professionals in an attempt to stay competitive. It would be misleading to assume that the suppliers of sub-contracted labour share the same values as the main contractor. Similarly, it would be unreasonable to assume that the suppliers of sub-contracted work share the same values as the architects or engineers: they do not. Discussing and sharing of values, with the aim of establishing some common project values, is crucial to the successful development and delivery of projects. It is also important to recognise that values and priorities change during the course of the project. Communication skills are essential to recognise and respond to the values of others and to align and reinforce the values of the project team.

Individuals and their organisations bring differing values and interests into the construction team and it is to be expected that differences of opinion emerge as different interests and values are developed and challenged. Hill (1995), Azam *et al*. (1998) and Thomas *et al*. (1998) claimed that the way in which an organisation is structured and operates, and hence its success, is determined by the communication practices employed. This applies equally to the temporary construction team and to the individual organisations that constitute the various supply chains. There is a strong bias in the construction management literature to the reporting of unsuccessful projects, despite the fact that many projects progress to a successful completion. The successful projects appear to be of less interest to researchers and tend to go unreported, thus making it difficult to compare successful and unsuccessful projects objectively to identify critical success factors. With moves to benchmark projects objectively, this bias may be addressed over time. Many projects experience problems (both small and large) during their life cycle, despite the best intentions of the actors involved. We believe that this may be partly due to insufficient attention to the manner in which organisations and individuals interact during the life of a project, in particular the failure to understand how the participants communicate and how they influence each other through communication. Organisations pay far too little attention to the communication skills of their staff and the interaction practices they employ. However, it is important to realise that this is a complex topic and models of best practice and solutions to communication problems are not easily developed.

Concern about the disaggregated nature of the construction sector has resulted in new (or rediscovered) approaches to the design and construction of buildings. Associated with this is the publication and promotion of new forms of contract, some of which aim to remove some of the organisational barriers and stimulate teamwork. Contracts are concerned with formal communication routes, ideal situations that few participants follow to the letter. Informal communication routes are required to make things run smoothly and are regularly used. Indeed, many actors may be quite ignorant of contractual arrangements until something goes wrong; they simply have a job to do. The response to the new contractual forms has not, however, been to change the way design and contracting organisations are structured, leading some to argue that little has changed under the surface (Cox and Ireland 2002). Although new contractual arrangements have been developed (and no doubt will continue to be developed), the construction sector still operates with different organisations (or different departments) dealing with design and construction. Regardless of the contractual arrangement actors have to communicate over organisational boundaries, and communication barriers still exist (Swaffield 1998). So, it may be necessary to examine successes within the current structural constraints and identify the type of communication practices that are associated with successful projects.

Professional and organisational cultures interact at boundaries within projects. Obvious cultural boundaries are the interfaces between client and brief taker, brief taker and design team, design team and contractor, and contractor and sub-contractors. Other more subtle boundaries, for example between architects and engineers, also exist. In many cases the interface is effective and projects progress well, delivering a building that satisfies both client and end users. Conversely, interaction difficulties are experienced and may quickly spiral out of control into a dispute. There is a need for effective communication on two levels, intra-organisational (within organisations) and inter-organisational (between organisations). Organisational boundaries within construction projects are constantly changing; individuals enter and leave the teams at certain stages (i.e. they have separate goals), the team changes in size and format (Gašparík 1999). Meetings, which are often used to transfer information between different organisations, are all too often performed at sub-optimal levels (Cummings 2000).

Language can, and frequently does, present challenges to the efficient management of construction projects. Professionals have developed specialist languages, using words that are specific to their professional background (Gameson 1992) to enable them to communicate specific facts and ideas quickly to fellow professionals (Emmitt and Gorse 2003). The construction process relies on the actors from various professional disciplines with different vocabularies and people easily fall in to the habit of using what others perceive to be jargon – a codified language that is difficult to understand by those lying outside the professional culture. The use of some of these terms may lead to confusion if not supported with sufficient background information and explanation. Additionally, some words will carry more meaning and information when used in a specific context and when exchanged with other actors who share the same language.

Workers from different parts of the United Kingdom use different words for the same thing and some dialects offer the potential for misunderstandings and confusion. Add in the increasing migration of construction workers in Europe and it is not uncommon to find construction sites with a multi-national workforce using their native tongue to communicate. The potential for misunderstanding is never too far from the surface in such environments, even where everyone's intention is to do a good job. The increase in the number of professional roles and market forces also means that many of the parties are competing for the same business while, somewhat paradoxically, also collaborating on projects. Thus, communication may not always be as open or as frequent as we might expect.

Relationships can be volatile and adversarial, making it difficult to form and thereafter maintain inter-organisational relationships (Hall 2001). To a certain extent initiatives such as partnering, strategic alliances and integrated supply chain management help to mitigate the effects of fragmentation, although, as with more traditional approaches there is still a heavy reliance on the ability of the team members to interact and communicate effectively. Thus, regardless of the approach adopted, the basic tenets of running a project remain the same, that is, we are reliant on getting the right people together for the right job. Competences and the development of competent practices are a key factor in the success or otherwise of the construction team. Research undertaken by Naoum and Mustapha (1995) and Charoenngam and Maqsood (2001) led them to conclude that the attributes and actions of key construction personnel strongly influenced the success or failure of the project. Recent research in Denmark into the composition of construction teams to facilitate communication within projects, and hence improve the success of the project, supports this view (Emmitt *et al.* 2004).

## Interaction during construction projects

Interactions during the early stages of a project are conducted in a relatively free and open atmosphere because the level of information in the system is relatively low and unstructured (Wallace 1987). Over time, the amount of information increases and formal and informal structures emerge, creating implicit rules of engagement. Goals and targets are set and the pressure to develop and exchange information increases. The number of actors contributing to the process and the amount of information are at their peak during the construction process, and it is here that the possibility for misunderstandings and conflict is highest. Morgan and Bowers (1995) review of research on stress and teamwork found that the pressure resulting from limited time and increased workload can affect the ability of team members to work together, leading to changes in communication patterns that may either impede or improve performance.

During construction the contractor's prime functions are to coordinate information and to allocate, manage and coordinate resources (e.g. the actions of specialist sub-contractors and numerous other participants). At this stage the roles and responsibilities of the architect and contractor, with regard to communication with specialist sub-contractors and consultants, may become complicated. Although drawings, specifications and schedules from the contract documents are often assumed to be complete and free from error, the reality is that they are not (Calvert *et al.* 1995; Harvey and Ashworth 1997; Gorse 2002). Much of the interaction during the construction phase is therefore concerned with the final development of detailed designs, changes to construction details and coordination and integration of the specialised components to ensure that the building can be assembled safely and within agreed parameters of time, cost and quality. A key objective for designers is to ensure that the information flow does not interrupt the construction process, but this is not always achieved in practice. Research conducted by Loosemore (1996a) found that information may be insufficient to perform a task and may result in a crisis. Indeed, it is probably unrealistic to expect the participants to know everything or be prepared for every eventuality. Thus, all actors need appropriate skills to be able to provide and extract the relevant information to stimulate and influence the behaviour necessary to deliver the project successfully.

Clarification of information and discussion of progress and problems is usually conducted within the forum of formally scheduled progress meetings and *ad hoc* meetings called to discuss a specific problem. When discussing complicated issues each actor will need to explain the situation in such a way that others (less familiar with the issues) may understand the context and hence be able to differentiate between what is important and what is not. Professionals may use different communication mechanisms, expressions and emotions to ensure that others pay greater attention to their message. Types of actions that ensure that other professionals' attention is focused on their interaction may be very useful in environments where large quantities of information are exchanged.

Open and supportive communication is conducive to building trust and facilitating interaction between construction team members. Loosemore's (1996a) social network research found that defensive behaviour developed during problem solving had adversely affected the ability of team members to work together. Ironically, when communication between team members is most needed, during times of uncertainty and crisis, relationships often break down and communication becomes difficult. In an investigation based on two construction projects, participants took part in group discussion sessions (quality circles) to identify the actions that had resulted in specific failures. Using the root cause analysis technique, Hall (2001) found that poor communications and misunderstandings resulted in failing inter-company relations. Hall (2001) claimed that open exchanges of information and sharing task responsibilities are essential for effective teamwork. Interaction that builds and maintains the fragile professional relationships necessary to accomplish tasks is fundamental to the project's success. It is important that research identifies how interaction can be used to strengthen and maintain relationships enabling participants to work on tasks effectively. Open exchanges may involve both supportive and critical comments. Open communication can be very effective, but if it is not managed in a sensitive and appropriate way for the context it may be destructive to relationships. All actors need to be equipped with the appropriate communication skills to ensure that matters that may cause conflict can be openly discussed while at the same time ensuring that the relationship is maintained and not damaged.

Negotiation, influence and persuasion have an important role to play within a business setting. Redmond and Gorse (2004) found that subtle differences in the negotiation approach had a considerable influence on the success of the meeting. Thus, it is necessary to develop communication skills and understand the effectiveness of negotiation approaches so that they can be adapted to suit the business or project context.

## Researching communication during live projects

An examination of communication practices adopted by the key construction team members operating on successful and unsuccessful projects may reveal different interaction traits. The first stage, in this research project, was to identify aspects of interaction that could be investigated and compared to previous research, thus helping to provide valuable insights into communication behaviour. Communication is a complex issue and undertaking a research project to better understand communication in construction teams posed some formidable challenges. The first issue to address is the characteristics of the communication acts, that is, what type of data are we trying to collect and why? Trying to analyse who says what to whom is, in our opinion, a little too detailed and too project specific to identify trends or to have wider application. To avoid overcomplicating the data through examination of each individual word and the subtle nuances they carry, it was felt that the emphasis at this stage would be better placed on the characteristics of interaction. Thus, the exchange of information, opinions, beliefs, emotions and other communication acts and their relationship to group performance is a fundamental concern.

Our research attempts to understand how team members gain control over their physical and social environment, using communication signals supported by task and emotional content. Communication within the construction team is considered from an interpersonal perspective, exploring the use of communication by various actors and the effect that various types of communication acts have on group interaction and behaviour. The research deals with contextual issues, such as the environment and the number and type of actors engaged in the communication. Observing all communication stimuli would prove impossible, so observations were limited to verbal interaction, while recognising that body language and facial expressions provided important information in helping to understand the meaning of the verbal messages. The physical observation of group interaction is considered important; it will assist the researcher's interpretation of messages and their meaning.

Without pre-empting the discussion of methodological issues it is important to identify a number of factors that influenced the focus of this research and the research method. At an early stage in the research, an exploratory investigation using postal questionnaires, interviews and case studies was undertaken. The aim was to investigate the perceived effectiveness of communication media and communication problems encountered by construction participants (see Emmitt and Gorse 2003 for a full description). The results found that face-to-face (interpersonal) communication was perceived to be the most effective communication medium. We also found that meetings, both formal and informal, were perceived to be beneficial to the successful completion of construction projects. Collecting data from informal (unstructured and unscheduled) meetings posed some methodological problems that were difficult to overcome. Prearranged meetings are, by their very nature, more amenable to systematic study, and so a decision was taken to collect data from site-based progress meetings.

An analysis of previous research indicated that relational communication (interaction that affects the development and maintenance of relationships) and task-based communication (interaction necessary to complete tasks) between individuals from different organisations appeared to be a core problem. Relational communication is used to engage in and manage conflict. Failure to balance relational and task-based exchanges could prevent group conflict from being properly managed. An uncontrollable escalation of conflict will threaten the team's ability to continue a positive dialogue necessary for the progression of the project. This led us to focus on live projects. An investigation of 'real' interpersonal and group communication behaviour between designers and managers appeared to have the potential to make a significant impact on the lack of knowledge and understanding of construction interaction.

Effective communication could have a major effect on project outcomes. The research reported here investigates whether there is an association between a number of project outputs and part of the project's communication process (management and design team meetings). By obtaining interaction data from construction project progress meetings, the communication behaviour was examined. This helped to provide a greater understanding of communication and interaction during the management and design team meetings. Contracting organisations supplied information about project cost (over- or under-budget), project completion time (earlier or later than the programme target) and the incidence of formal conflict between the main actors. Information on the ability of the contractor's representatives to repeatedly deliver projects on time and within budget without contractual dispute was also made available to us. Combined, the data enabled patterns of interpersonal communication behaviour for the successful and less successful representatives to be identified and evaluated.

The first research objective is to determine whether:

1.  *The management and design meeting has a characteristic model of interaction, being different from interaction profiles found in other contexts.*

In relation to the first research objective, the research sought to test the following hypotheses that:

2.  *The interaction patterns of group meetings associated with successful project outcomes are significantly different from the interaction patterns of group meetings that are associated with unsuccessful project outcomes.*
3.  *The contractor's representatives considered to be most effective exhibit significantly different interaction patterns from the contractor's representatives perceived to be less effective.*

Testing these hypotheses had the potential to make a significant contribution towards understanding the nature of professional interaction and its association with project outcomes. The investigation of group interaction offers the potential to explore conflict and interaction related to its use and management. Issues of specialist participation, dominance and methods of influencing group interaction will be identified and discussed within the above context. Our aim was to provide:

-   a greater understanding of group communication between management and design professionals;
-   a better understanding of the trends, patterns and norms that emerge during interaction and determine whether they are different or similar to those found in other contexts;
-   information on the traits of group communication behaviour associated with successful projects (completed within cost and project duration and no major disputes) and those associated with unsuccessful projects; and
-   a greater understanding of communication traits of the contractor's representatives who have repeatedly delivered successful projects.

The book is structured to provide information to ground and support the final conclusions and recommendations. Building on this introductory chapter, the research into communication within construction processes is reviewed in Chapter 2. This leads into a discussion of group interaction research and its applicability to the construction sector (Chapters 3 and 4). A crucial consideration for construction communication researchers is the ability to select an appropriate data collection method from live projects and this is explored in Chapter 5, together with a description of the method used for the research presented in this book. Results of the data collection and data analysis are presented in Chapters 6, 7 and 8, which describe the case study results, team interaction characteristics and project outcomes respectively. Discussion of the research findings can be found in Chapter 9. By way of conclusion to the book, a number of practical considerations for practitioners and researchers are proposed in the final chapter (Chapter 10).

Being the first study of its kind in the field of construction we would like to encourage others to build on our data, thus helping to provide a sound basis for evidence-based communication practices. We also hope that we help to raise awareness of important, yet largely ignored, aspects of interaction common to all construction projects. Our hope is that by presenting the work in a logical and clear manner practitioners can reflect on the issues raised in this book and apply appropriate frameworks and working methods to facilitate effective communication in their construction project teams.

## Further reading

Dainty, A., Moore, D. and Murray, M. (2006) *Communication in Construction: Theory and Practice.* London: Taylor and Francis.

Emmitt, S. and Gorse, C.A. (2003) *Construction Communication.* Oxford: Blackwell Publishing.

Fryer, B., Egbu, C., Ellis, R. and Gorse, C.A. (2004) *The Practice of Construction Management*, 3rd edn. Oxford: Blackwell Publishing.

Moore, D. (2002) *Project Management: Designing Effective Organisational Structures in Construction.* Oxford: Blackwell Publishing.

# 2 Construction communication research

Research into group communication has shown that social interaction is essential for building and maintaining relationships and that task-based interaction is necessary to accomplish the group goal (Keyton 1999). Key to project success is the interaction practices that help to establish and then maintain the inter-organisational relationships necessary to accomplish the construction team's goals. The nature of the relationship established through communication will affect the actor's congruent understanding and hence the actor's ability to use information to aid decision-making (Gorse and Emmitt 1998a). The relational qualities necessary to deliver a functional building can only be achieved through effective social and task-oriented communication. Improvements in the construction process will only be achieved if we start to understand how we use communication to build and maintain the relationships necessary to undertake multidisciplinary tasks. The success of construction projects seems to be highly dependent on relational and task-based communication. In a construction context these two elements of communication are interdependent. Communication and the social context in which it occurs are interwoven and research must consider both factors (Gudykunst 1986).

One of the first findings of group interaction research was that groups exhibit patterns of communication that are directly related to their structure and context (Bales 1950). Some of the factors that have affected the group and individual communication patterns include: reasons for the group meeting; maturity of the group's participants; the participants' relationship with each other (Bales 1950); their professional goals, roles and experience (Gameson 1992); organisational relationships (Pietroforte 1992); and experience of the information and communication technologies used (Abadi 2005; den Otter 2005). Combined, these factors have implications for team assembly and for research into team communication. A fundamental requirement is to clearly define the environment in which the research is taking place, providing contextual information relevant to the research problem (Ragin and Becker 1992; Hughes 1994; Abadi 2005). It is to be expected that research projects undertaken at different times, in different contexts and with different participants may produce different results. There are, however, a number of factors common to construction projects that enable some comparisons to be drawn. These relate to group development within the temporary project team and the extent and nature of the actors' interaction during the project. Some of the characteristics and patterns of communication in a construction context, and indeed other contexts, may remain constant across projects; others may be specific to the context and/or the actors involved.

Communication research attempts to provide theories that explain a situation or event (Albrecht 1997). Unlike physical or chemical science, the social sciences are not exact sciences and by virtue of this many of the theories will not apply every time in all situations. The study of communication in the construction industry is applied research; it attempts to offer practical advice and explanation of real situations. *Communication research* has been defined by Hargie (1992) as *the scientific study of the production, processing and effects of signal symbol systems used by humans to send and receive messages*. This view presents three main dimensions (Hargie *et al.* 1999):

- Communication is capable of scientific categorisation; data from the process can be recorded and is open to interpretation, classification, measurement, analysis, evaluation and improvement.
- The study centres on how messages manifest are produced, processed or delivered and what effect they have on those who receive them.
- The importance of the signs and symbols is highlighted. Interpersonal behaviour serves a communicative function, being judged on the basis of verbal content and non-verbal signals. Understanding the signals and identifying the clues that give meaning to the signal is an important aspect of the communication process.

Hargie *et al.* (1999) argue that if we are to improve communication, we must first understand the communication process: but this is not an easy task (Fiske 1990). Even though we constantly engage in communication, in organisational and private settings, it is an activity that few can define satisfactorily. A review undertaken by Dance and Larson (1972) found 126 definitions of communication, since which many more have been formulated. Definitions are usually specific to the context in which communication occurs, and/or closely related to the purpose or field of research. There are, however, some underlying similarities. The most common features or sentiments of these definitions include: communication is the transfer of meaning; a response of an organism to a stimulus; the process through which people make sense of things and share this with others; the transmission of information, ideas and emotions; an act initiated by individuals in which he or she seeks to discriminate and organise cues to orientate themselves in their environment; and a mechanism to influence the environment (Trenholm and Jensen 1995). Burgoon *et al.* (1994) provide a simpler and more pragmatic description of communication as a tool by which people may gain some control over their social and physical environment.

It is not so critical that the definitions of communication differ; rather it is important that the attributes of communication that are to be researched are adequately defined. Classification schemes are often used to assist in the analytical distinction, helping to identify the unit of analysis (Bowen 1993). One of the most prominent distinctions made between different types of communication is that of level. In construction management literature, differentiation between levels tends to be made in terms of operation and actions, from the micro-level (individual) to the meso-level (group) and then to macro-level (project); however, it is more common in communication research to define levels as shown in Table 2.1. Here, the process and the people involved are discussed in five levels, from intrapersonal to interpersonal, group, multigroup and mass communication. Our research will focus on interpersonal communication and group interaction because of its importance to building and maintaining effective groups and teams.

*Table 2.1* Levels of communication (From Gorse 2002)

| *Process* | *People involved* |
| --- | --- |
| *Intrapersonal communication* | |
| Internal communication process (cognition); includes the manifestation of information in the brain, which is understandable and has some relevance. Intrapersonal communication would also include the knowledge that another person is able to process the information sent (relevant information). The way others behave or react to messages stimulates conscious and subconscious processing in the mind of the observer. Attempting to understand situations around us is natural. It is important to note that we are affected by the information received. | Only one person is involved. It is the thought process of one person either when they are alone or when communicating with others. If processed at a subconscious level, it may not be considered a thought process. Intracommunication may be used when one person makes a decision alone, or simply observes communication or the environment around them. As there is often only one person involved some scholars do not view intrapersonal communication as a communication process. |
| *Interpersonal communication* | |
| Face-to-face communication between people enables individuals to establish and maintain relationships. It involves the transfer of signals, clues and messages that manifest in both parties to the communication act. Intrapersonal and interpersonal communication functions enable information to be processed and joint decisions to be made. Each party sends out signals, attempting to affect the cognitive environment of the other, while both parties process clues from the signals to help develop understanding and engage in conversation. Most face-to-face interaction is processed in the subconscious mind; thinking out every word before being spoken would result in slow conversations. | Generally this involves two people (dyad), more than two people may be considered to be a group. Although some scholars do not differentiate between two people and a group, there is a significant difference. With communication between two people, the message is only intended for one receiver. When a message is broadly directed at a group it is not always known whether the message is aimed at one or more persons within the group, whereas in one-to-one conversations there is no doubt who the message is intended for. With certain comments this polarity significantly increases the intensity of the message. |

*Table 2.1* (Continued)

| Process | People involved |
|---|---|
| *Group communication* | |
| Messages are sent to the group, they may be presented in a way that addresses the whole group or individuals within the group. Messages may be interpreted and processed by individuals within the group in different ways. Members within the group may use other members' reactions and responses to understand the message, its meaning and intent. Terminology and language may be specific to the group. Each group will have its own culture and norms and may act and react differently. | Involves more than two people but limited to a single group of people. Communicators may address the whole group or individuals within the group. When personal conversations take place within a group, others will be aware that the conversation is taking place, even if the content of the message is not heard or understood. In small groups members are extremely vigilant and aware of the behaviour, actions and reactions of others. |
| *Multigroup communication* | |
| A person or group communicates a message to a number of different groups or subgroups, the response to the message may be different depending the groups' motivations and norms. Subgroups may react differently to messages. | Although communication of this nature targets a number of groups or subgroups, there is an element that the messages are largely contained within the specific groups, for example departments within an organisation. |
| *Mass communication* | |
| Messages are sent through, media, radio, television and newspapers to large audiences. Individuals and groups of people receiving the message may attach different meanings to it depending on their culture and norms. The message is often designed to attract the attention of the largest audience. Such messages often carry information that stimulates emotional responses. | Offers some control of who and how many receive the message. Groups can be targeted for example television viewing at a particular time – advertising to children, for example on Saturday mornings between children's programmes. New communication systems enable TV, newspapers and mail shots to target specific regions. Professional journals are used to send information to their profession. |

Interpersonal communication occurs when two or more people are engaged in a form of communicative exchange (Kreps 1989). Reinard's (1997) work on communication identified interpersonal communication as interactions occurring in one-to-one and small group situations. Price (1996) suggests that for group communication to occur, the collection of individuals forming the group needs to share common attributes, goals or interest or at least have some common values or norms of behaviour. Kreps' (1989) definition of small group communication has some similarities to that proposed by Price. For Kreps, small group communication occurs when three or more people interact in an attempt to adapt to their environment and achieve commonly recognised goals. In the context of the construction project it is the interpersonal and group interactions between members of the management and design team that are potentially of greatest interest. It is here that the gulf is greatest between the values of the designers and the constructors.

Individuals may contribute to group activities for many different reasons, for example social or task based. The context does not always set the agenda for being involved in the group, although they are normally closely related. Involvement in work groups may have greater social attraction than the task they are intended for. Conversely, people may join a group to sabotage the group's goals (which is not always evident until it is too late). Groups are often said to have a common goal, which is little misleading. It is important to recognise that the group goal may be either social or task based, and this is both conscious and subconscious. Normally groups have a reason, motive or driver, behind their communication. At the most basic level, people form groups to interact and socialise, gaining comfort and belonging from the group. At much higher levels, people gain considerable power from the group and are able to accomplish things that would be difficult or impossible in isolation.

## Group development within the temporary project team

It is common, and perhaps a little too convenient, to consider that the participants involved in the construction project operate as a single organisation. It should be recognised that the organisation is transient, only lasting for the duration of the project, involving a large number of sub-contractors (Cherns and Bryant 1984; Hill 1995). Thus, the construction team is in effect a loosely coupled arrangement of interdependent teams, groups and individuals all contributing in different ways and at different times to the realisation of specific tasks. In the vast majority of construction projects the participants are brought together to work on one project only. On completion of the project, or more accurately on completion of a participant's particular work package, the relationship between the individual and the project stops. This means that, with the exception of large and repetitive projects, it is common for the project team to be composed of actors different to the previous one. This is often true even where the same organisations are involved, simply because different individuals within the organisation have been assigned to the project according to internal workload commitments. Thus, communication is required to support industrial relations and hence to provide the means by which teams can develop quickly and effectively.

The small teams, groups and individuals that constitute the construction team are dominant at different stages in the project. Thus, although interdependent upon the work of others, many of the participants will not meet other participants during their period of involvement with the project. Relationships form, evolve and disband throughout the life cycle of the project. This is an important point to make, because newly formed groups exhibit characteristics different to established groups (explored in further detail in Chapter 3). In a newly formed group, members must simultaneously find their place within that group and socialise with one another (Anderson *et al.* 1999). Each construction project embraces a new grouping of actors who must attempt to develop social relationships that support the project. Research on group interaction and socialisation processes found that in new groups the exchanges present a two-way process of social influence, changing other's ideas and opinions through communication (Anderson *et al.* 1999). It is claimed that the norms that govern group members come into force as each of the actor's objectives become apparent (Lawson 1970, 1972; Derbyshire 1972). The group norms regulate group behaviour. Wallace (1987) found that as groups developed, the importance of individual professional roles would be recognised; the group regulatory forces are then used to control interaction of individual members. In construction, as the tasks change from feasibility to design and then construction, the power, importance and involvement of the professionals will change. The challenge for project managers is to try and lead this dynamic process. As the project develops and the specific demands of the situation changes the actors with the most relevant skills become more influential and powerful. Those with relevant skills emerge or are nominated by other members of the group to be the most dominant for a particular period of time, during which they are central to the information processing and decision-making because of their expertise. Most construction projects and the teams assembled to deliver them are new ventures and regardless of the expertise and experience of the actors they are often presented with unfamiliar situations.

It could be argued that designers and managers are often presented with unfamiliar situations wherein they do not know how to act. Any social situation is a sort of reality agreed upon by those participating in it, or more exactly those who define the situation (Scheerhorn and Geist 1997). Everyone who enters a situation does so with preconceived definitions of what is expected of him or her and the other participants (Bentley 1994). Such beliefs, which include expected interactions, are established from experience of previous groups to which the individuals had belonged (Scheerhorn and Geist 1997). Thus, each situation confronts the participant with specific expectations and demands. Such circumstances generally work because most of the time our perceptions and expectations of important situations coincide approximately (Berger 1986). Culture, society and individual power play an important part in the norms that govern they way social groups act. The same aspects will also influence the behaviour of professional groups, but the group may also draw on those members with nominated roles, perceived expertise, skill and experience to determine who is allowed to interact, make decisions and play lead roles. However, some people, regardless of professional skill or experience, can use communication techniques to exert influence enabling them to gain power over elements of group behaviour and decision-making.

Not all teams are totally new, and group norms and behaviour may be influenced by previous relationships and experience of that relationship. Although the project is often a one-off and teams will disband, some groups will maintain relationships and use that relationship in future projects. For example, if a sub-contractor is required for a new project a contractor may be minded to use those worked with in the past, especially if the relationship and performance was good. Clients, developers and consultants usually have a number of preferred contacts (individuals within organisations) that they will use if given the freedom to do so. Alternatively, parties may contact those that they have previously worked with for advice or help without entering into a contractual arrangement. Working with known actors through informal strategic alliances or formal strategic partnering may save time and improve knowledge transfer because relationships have developed, a level of trust exists, some formalities have been dispensed with and the skills are known (and to some degree have been tested).

## Communicating to achieve project objectives

All actors need to collaborate, share, collate and integrate significant amounts of information and knowledge to realise project objectives (Pietroforte 1992; Oxman 1995). To do this well requires a combination of good management, committed people, supporting ICT networks and the opportunity to regularly communicate on a face-to-face basis. Research shows that for addressing contentious issues, problem solving, conflict resolution and building relationships, face-to-face communication is crucial (Abadi 2005).

The quicker information can be accurately produced and understood by those dealing with it the better. Much attention has been given to the transfer of information within project environments, since the timely exchange of information is a key factor for effective co-ordination and hence control of project information. A considerable amount of information can be processed at remote locations, for example in the office where all the tools, equipment and information are handled, and information produced can be easily sent via email or web-based technology. If actors are working in different countries with different time zones (through some form of collaborative venture or through outsourcing work offshore), the information that is complete at the end of the day can be sent to other offices so that the other parties can continue working on the project. Use of technology in this way allows information to be processed quickly. When the information is simple, easy to understand and non-contentious, face-to-face meetings may be unnecessary and wasteful.

Asynchronous communication (where parties do not communicate at the same time, for example by email where messages are sent and then responded to) can be quick, but in face-to-face communication, where information is sent and received at the same time via different clues, signals and expressions, so much information is processed at one time that the ability to resolve problems and issues is substantially increased. It is clear that different forms of communication have different uses, advantages and limitations.

Information is required to enable the processes to proceed as planned and also to control change as and when it occurs. Growing use of intra-webs and project-extranets has made a significant impact on the way in which information is exchanged between project actors. It would be prudent, however, to remember that people produce information, and we often make mistakes, especially when working under pressure. Subsequently some form of interaction between the sender and receiver of information is required to resolve identified discrepancies.

Drawing on a vast body of group communication research, Keyton (1999) claims that communication in any group has social and task dimensions. Task roles are those that determine the selection and definition of common goals, and the workings towards solutions to those goals, whereas socio-emotional roles focus on the development and maintenance of personal relationships among group members. Higgin and Jessop's (1965) pilot study found that interaction of professionals during meetings was used to build and maintain the relationships necessary to accomplish tasks, an early realisation of the importance of meetings. The communication traits of key individuals and the interaction practices of core groups were also seen to be important for problem solving during the construction phase. Problems that occur during the construction phase require the involvement of design and management professionals to resolve them. As problems develop, key members of the design team and contracting organisation will need to use effective communication to co-ordinate the available information and hence allow decisions to be taken and the project to proceed (Yeo 1993; Hugill 1998; Macmillan 2001).

Interpersonal relationships are developed through the behaviours of individuals. We will respond depending on how others behave towards us. Even when the period of interaction is short, there will be signals sent that will influence the behaviour during the communication act. Signals transmitted during communication help to provide feedback on the relationship. Signals of this nature are known as *metacommunicative messages* (Kreps 1989; Lesch 1994). Typical providers of metacommunication include remarks or gestures that are directly related to what is being said. Such signals help us to understand why something was said and why it is important to the person making the statement. Personal reactions by the receiver, in the form of metacommunication, would also send clues to the sender if the message were not properly understood. Relational communication signals of this type are used in the development of implicit contracts, helping to understand how the other person feels, recognise mutual expectations, and identify what is agreed and what is not (Kreps 1989; Lesch 1994). As relationships develop, implicit contracts are formed which enable strong relationships to be formed. Metacommunication provides us with information on how to act together to accomplish tasks (Kreps 1989). Failing to use and process metacommunication may mean that issues of potential disagreement are not recognised and conflicts not managed. Metacommunication will provide information on potential areas of conflict and agreement, both being important to the development of an amicable relationship.

When the construction project starts, the architect and main contractor (acting as project managers for design and construction respectively) are, arguably, the most important communication nodes. The flow of information during the construction phase is channelled primarily through the architect and main contractor. The dominance of the contractor over the architect or vice versa will, to some extent, depend upon the type of contractual arrangement entered into, and their relationship with the client. Members of other groups that are party to the project (e.g. sub-contractors) are dependent on these intermediaries for their information. Thus, both the architect and main contractor occupy powerful positions when controlling project information and communications. Successful and efficient construction projects depend on the highest degree of co-operation, at every stage and between every level, but none more so than between the two functional partners, the architect and contractor (Calvert *et al.* 1995; Byrd 1998; Emmitt 1999).

Gardiner and Simmons (1992) have suggested that personal differences between architects and construction managers can lead to conflict, but conversely can also be used to prevent conflict. It would seem that historical separation in education and development of strong professional identity has led to dissimilar perceptions of social status and role definition. Stereotypical images of architects and construction managers are easy to find, and many actors have quite strong perceptions of how certain professional types should act. Pietroforte (1997) found a dislocation between the roles and rules of standard contracts and the actions of the professionals, noting that successful contracts are completed through co-operation, informal roles and rules, which both complement and circumvent standard contracts. These findings tend to indicate that professionals may not rigidly adhere to the prescribed procedures. Wallace's (1987) research supports this view, concluding that interaction patterns of professionals are a function of group process as opposed to administrative factors imposed by contracts and organisation. Wallace (1987) found that although professional status would provide participation rights during early interaction, the problems faced by the team and group control mechanisms were a greater predictor of who participated in the group's tasks during later stages. The effects of status on communication and decision-making have been found to be different for permanent and temporary groups (Hare 1976). In permanent groups (groups that have been together for a long time), the higher status members have a greater influence on the decision-making process than higher status members in *ad hoc* or temporary groups (Hare 1976).

## Construction communication research

A small number of doctoral studies have tackled aspects of communication in construction. The most relevant of these include work by Kreiner (1976), Wallace (1987), Gameson (1992), Lawson (1970), Pietroforte (1992), Bowen (1993), Loosemore (1996a), Emmitt (1997), Larsson (1992), Hugill (2001), Gorse (2002), Abadi (2005) and den Otter (2005). In these works the importance of interpersonal communication as a 'glue' to keep projects moving forward is evident. Some of these researchers deal with discrete aspects of the construction process, such as interaction in the feasibility stage (Gameson) whereas others (Larsson and Emmitt) have investigated communication between specific actors, manufacturers and contractors or manufacturers and architects respectively. Kreiner, Pietroforte, Loosemore, Hugill, and Gorse collected data on distinct events and situations that occurred during the construction phases. Abadi and den Otter concentrated on project web tools to facilitate communication.

A few points emerge from the studies of communication during the construction process. There is a lack of research that observes interaction of the construction manager and other key professionals during site-based meetings. The research was conducted in different countries, for example Kreiner's research was conducted in Denmark, Pietroforte's in the United States and Bowen's in South Africa. The findings are useful, but care is required in drawing comparisons because cultural, legal and contractual differences exist. From the findings of the research it is clear that interaction between actors is used to resolve professional differences, to clarify conflicting requirements, to overcome crises and also to circumvent formal communication routes dictated by construction contracts. Face-to-face interaction is still considered the preferred method for resolving problems and contentious issues, yet there is little information on the behaviour and practices of professionals during face-to-face meetings.

### Communication research during the feasibility stage

Gameson (1992) investigated interaction between potential building clients and construction professionals. Gameson arranged 44 meetings during which experienced and inexperienced clients interviewed different types of construction professionals, on a one-to-one basis, to discuss how a potential project could be managed. The interaction patterns and behaviour of different types of clients with different construction professionals were examined to determine whether the clients' experience and the construction consultants' professional background affected interaction. Although Gameson's research observed face-to-face interaction, this was staged in a laboratory setting (experimental direct observation). He found that different types of professionals and clients exhibited different interaction patterns depending on their profession and experience. Data, which were collected from an initial meeting, showed that construction professionals from the same profession tended to use some words more than other professionals would. For example, quantity surveyors made greater use of words associated with cost and legal issues, such topics being more aligned with their educational and professional background. Gameson also found that clients and professionals with more experience tended to dominate communication during the meetings, directing and questioning the other parties. Inexperienced clients would listen to the advice given, without exerting the same degree of questioning on the other party. The degree that a person is able to question another's statement may be linked to the degree of knowledge the questioner has in that particular area.

### Communication research during the pre-contract phase

Wallace (1987) observed the interaction of four actors (architect, client, consulting engineer and quantity surveyor) and the influence of design team communication content upon decision-making processes at the pre-contract design stage. Interaction was recorded as it occurred (using natural direct observation). Two types of observation were conducted. One longitudinal study of a series of group meetings was undertaken to provide a continuous picture of the group's evolution throughout the pre-contract design stage. The observations of the longitudinal study were supported by cross-section studies collected from 16 projects, providing observations of the group at brief periods in the design stage. The intermediate observations were undertaken at the outline proposal, scheme design, detailed design and production information stages. Interviews were also conducted with the subjects to provide qualitative support for observations.

Through the examination of group interaction during this stage, Wallace found that decisions made by the architect were a function of an evolving group development process. Changes in the content and type of group interaction emerged as the project developed. The architect was prominent during early stages, where he or she was initially perceived as team leader. During the initial meetings, the architect was found to have considerable freedom and interaction flexibility, and thus imposed their ideas on the group. As the design information developed through multidisciplinary group interaction, new goals were discovered and these led to changes in the group's interaction patterns. The interactions of the architect in the design team were less pronounced during the middle stages of the pre-contract design process. At this stage, interaction changes were evident, with the design requirements being imposed by other members of the team on the architect.

As the project design developed and became more complex, the architect was forced to draw on the practical experience of the other group members to resolve design problems. This resulted in changes in the group norms, which allowed other design team members to challenge the authority of the architect. Additional cost information presented to the group resulted in the formation of conflict. At this stage the architect was said to progressively disassociate himself or herself with the group process and the new group goals (which focused on costs). The increase in conflict led to the formation of sub-coalitions. Wallace's findings show that as the design develops and different specialists become more involved, ambiguities and incompatible approaches surface and conflict emerges within the team. The evaluation and selection of design information is a result of relational interaction, a process that determines who influences the integration of information and hence the final design.

Wallace's results suggest that when the subjects believed they were less influential in the group process they were less assertive. When less assertive they made fewer attacks on proposals and tended to be more defensive. Communication acts such as asking more questions and giving opinions and information were considered to be indicative of uncertainty, making subjects less influential. Subjects with greater influence also asked questions to extract information and give direction. Perceptions of influence were based on observations by the researcher and subjects. Where members of the group were found to agree with others on a particular issue, or wished to defend their position against other influential members, a subgroup or coalitions would form. The development of the coalition was assisted by supportive exchanges, fewer attacks and less expression of dissatisfaction. Wallace's results are useful, but unfortunately his thesis fails to provide detailed description for each communication act making interpretation of other results problematic.

Bowen's (1993) cost communication questionnaire, based on 124 responses from clients, 100 responses from architects and 99 responses from quantity surveyors, gathered perceptions on communication practices necessary to build 'price messages'. Although covering aspects of communication, Bowen's research is concerned with cost data and price models; as such the findings offer limited information to support the study of group interaction during the construction phase.

Abadi's (2005) research focuses on the issues and problems that face participants attempting to communicate project information. A major consideration of the research is the context in which most construction teams operate. Due to the rather fragmented nature of the construction industry, most teams are made up of designers and contractors, who are employed by different companies and who work in locations remote from the construction site. Abadi defines this as a *virtual team*. Abadi investigated the use of different communication media and their use at different phases within the construction project.

Questionnaire results suggested that certain aspects of construction were more dependent on face-to-face meetings when the team felt the need to work together. During such stages, information and communication technologies, which were used for remote working, were seen to be less of an advantage. Design was one aspect of construction where participants felt that they were able to work remotely using modern technology to communicate design information. The research, which focused on the design stage, suggests that ICT is playing an increasingly important role in the construction project, but there are many barriers to its effective use. Most actors still consider that face-to-face meetings are essential for building relationships, addressing contentious issues, resolving problems and managing conflict. ICT was being used to develop and provide design information, ask questions, request information, send information and store information. When working in remote locations, significant benefits were identified with the use of information technology over traditional paper-based systems; however, the practices of a few members within the team could often undermine such benefits. Each company within the virtual team had a different communication culture. Due to the different levels of training, experience and skills, specifically with the use of ICT, some companies would not use, nor learn how to use, some of the technologies that the management team had set up for the project. Some professionals exploited different communication techniques and technologies to improve their efficiency while others, who did not comply with systems in place, significantly reduced the efficiencies gained. Selecting participants for construction projects is currently determined by their cost and construction skills. If Internet or web-based systems are to become a significant part of construction communication then clearly participants must also have the skills to use these systems. Abadi's research also underlines the importance of face-to-face communication for developing relationships and resolving problems. Regardless of phase, it is clear that different communication media have their own strengths and if employed correctly can enable more relevant and efficient communication. Many of the communication methods complement each other, but thought needs to be given on how they can be used effectively.

## Communication research during the construction phase

Kreiner (1976) studied the social relationships on Danish construction sites, focusing on interaction during construction site progress meetings. Drawing on the work of Goffman (1974) he described the site meeting as a 'ceremonial event', where actors met to act out relationships in accordance with a fixed agenda. Observation of the site meetings revealed relatively stable relationships and harmonious debate with little exchange of information, a situation he described as 'uneventful'. Some participants did not contribute to the debate. The conclusion was that participants attended the site meetings to fulfil their contractual obligations. The findings helped to illustrate the importance of the site manager and in particular the ceremonial staging of the site manager in authority. Although Kreiner was rather dismissive of site meetings as a place to exchange information he found that the co-presence of a group of people allowed for misunderstandings and ambiguity to be removed, agreements reached and plans accepted, that is it provided the forum for coordination and problem resolution. The overall conclusion of the work was that more attention to the nature of interaction in organisational contexts was required. Especially the manner in which consensus is reached – consensus being a prerequisite for concerted (organised) action.

Using a longitudinal study of the design of a granite veneered façade, Pietroforte (1992) examined both formal and informal information generated and communicated during the execution of the design, purchase, engineering and testing of the system. This longitudinal study traced and observed the interactions of a major specialist sub-contractor, the cladding sub-contractor. Pietroforte's (1992) work identified a number of differences in the organisational structures of main contractors and façade sub-contractors, as well as the differences in the way construction projects were organised. This would suggest that while construction contracts identify responsibilities between parties, the organisational and operational arrangements that emerge are specific to the situation and the organisations engaged in the project, rather than being directly attributable to the contractual arrangement. This factor makes the process of identifying real responsibilities and relationships between the parties rather ambiguous, suggesting that the models of organisational structure do not always apply. Similarly, Loosemore (1996a) identified that social adjustment, behavioural instability and power struggles were some of the main factors that affected the resolution of unplanned events, rather than contractual procedures.

Loosemore (1996a) investigated the patterns of communication and social structures that develop as a response to a crisis during the construction phase. The actors observed included clients, estates department representatives, project managers, architects, clients' quantity surveyors, contractors' project managers and contractors' quantity surveyors. The research examined interaction that developed following a crisis and used social network analysis to examine data gathered through participants' reports. Loosemore found certain patterns in communication behaviour and interaction. Communication networks emerged out of a need for information to overcome a crisis. Construction crises resulted in conflict that discouraged collective responsibility and reduced the effectiveness of the communication network. The resulting conflict also generated latent tension that continued to appear cyclically throughout the phase. The crisis had a detrimental effect on communication. Paradoxically, it was also found that the same crisis that created a division in teams could also present opportunities for increased cohesion, harmony and efficiency. Parties could demonstrate commitment and sensitivity to the other professionals' needs, increasing cohesion and strengthening mutual trust within the project team. Where cohesive teams emerged, the efficiency with which unexpected sub-crises could be dealt with increased. Lee (1997) and Cline (1994) also found other benefits from conflict. They observed that disagreement can be productive, helping to expose alternative views and reduce the potential of 'groupthink'. Loosemore (1996a) noted that cohesion following a crisis can be used to resolve problems. Although Loosemore claims that some episodes of conflict strengthen relationships, the exact nature of such conflict was not reported.

Hugill's (2001) work on a team meeting from one project was mainly concerned with the method used to study the meeting. He noted that the method used, conversational analysis, had been particularly useful for examining short episodes of group behaviour, but was restricted in its ability to examine multiple meetings due to the amount of time taken to analyse data in this way. Discourse analysis, which could fall within the scope of this work, and conversation analysis have been differentiated on methodological grounds (Potter and Wetherell 1987). While discourse analysis is a broad term covering all forms of interaction, conversation analysis is restricted to verbatim transcripts of interaction rather than on field notes. Conversation analysis is considered to take a more naturalistic approach, it is focused on people's talk and how contributions are meshed together and the way different types of actions – blaming, excuses and greetings etc – are produced and managed. Conversation analysis may be considered a part of ethnomethodology. Ethnomethodology is concerned with the study of people and their methods for making sense of everyday life; analysis of conversation using such an approach is demanding. Due to the amount of detail extracted, studies are often limited to short episodes. The aim of ethnomethodology is to establish an understanding of the situation rather than testing a hypothesis. Most of the time the data is unstructured. Analysis of the data involves interpretations of the meaning and function of human actions, the product of which normally takes the form of descriptions and explanations (Atkinson and Hammersley 2001). Such detailed and intricate methods restricted Hugill's (2001) report to the analysis of one hour of the meeting's discussion. Although his observations were of meetings undertaken during the construction phase, his report presents a very brief snapshot of this period. Due to the length of interaction reported, the findings offer limited information. Such observations provide insight into human inter-action and behaviour, but to uncover trends and practices that affect the project many meetings need to be researched and the data collated. It is unlikely that an individual could undertake research of multiple meetings using ethnometh-odology because of the amount of time required. Indeed, a team of researchers would also find it difficult. It is more likely that the small episodes of human behaviour, recorded using ethnomethodology, will develop over time, as different researchers examine the aspect of behaviour that they are interested in. As the studies emerge, patterns of behaviour may be identified and such behaviour may be linked to successful or unsuccessful interaction. The details of data collected using this method may be important; however, the method lacks focus and often means that resources may be wasted investigating aspects that have little impact on the group behaviour. Ethnomethodology is useful, to examine the detail of specific aspects of human behaviour, but research into group behaviour, within the construction context, is at such an early stage in its development that we have little understanding of the general patterns of behaviour.

## Professional engagement and procurement

Given the amount of attention paid to procurement routes and contractual arrange-ments, it may appear that the type of procurement route adopted would have a major influence on communication. Practitioners are constantly urged to use a particular approach to procurement because it is deemed (for a while) to be better than other approaches. Both research findings and experience from practice would tend to confirm that one approach is no better or worse than another, merely more suited to a particular set of circumstances at a particular point in time (Emmitt and Gorse 2003). Procurement is primarily concerned with the control of resources (risks associated with those resources) and, one might assume, communications. However, Pietroforte (1992, 1997), Hill (1995) and Loosemore (1996a) have all questioned the extent that formal contracts control the communicative behaviour of those participating, especially when managing events, resolving problems and dealing with a crisis. Their work helped to highlight the importance of informal communication routes.

A number of procurement systems have been developed. The four main forms include traditional, design and build, design and manage, and management contracts (NEDO 1985). Although traditional contracts are still the most popular procurement method, there has been a reduction in the number of traditional contracts and a growth in design and build procurement methods (Gameson 1992; Langford *et al.* 1995). In the United Kingdom, the two most common procure-ment methods are general (traditional) and design and build. Both procurements systems were encountered during this research. Although each procurement path offers certain advantages and disadvantages, the same actors are involved; it is only the contractual relationships that vary (Gameson 1992). Thus, issues concerning communication and the building of a collaborative culture within the project team are omnipresent regardless of the type of contract used.

Construction project management publications identify communication structures related to specific contract arrangements. While helping to illustrate assumed (formal) relationships such work may be misleading. Research on professional interaction in live construction projects has found that communication does not always follow formal channels (Pietroforte 1992; Hill 1995; Loosemore 1996a). Formal communication channels may be found to be ineffective, and the procedures and contracts that govern and determine their structure are insufficient to deal with all circumstances (Hill 1995). Indeed, as already discussed, the professionals will pursue social structures that benefit the individual organisation (Loosemore 1996a), that is they will use informal communication routes to achieve their objectives. Even when management structures and systems are established and training provided, those not skilled or unwilling to use the system may persist in their own individual methods, forcing others to conform to practice outside the defined systems (Abadi 2005).

Pietroforte (1992) found that interaction between professionals was often different from that stated in the contract. A critical comparison between the 'assumed' (that written in contract documents and procedures) and what actually happened (observed behaviour) revealed a number of differences. He also concluded that where formal or administrative provision does not exist, or is incomplete, the formal provision is integrated by the development of co-operative attitudes. He suggests that much of the construction process is based on informal relationships and roles, especially when dealing with interdependent functional problems and transfer of small sets of uncertain or uncodified information. While status and elected position would enable certain professionals to participate more than others during initial meetings, it is the group's regulatory framework (norms) that exerts an influence on the individuals, dictating their roles within the team during subsequent interaction.

Regardless of construction contract, the professionals who make up the management and design teams receive and process much of the information generated by other groups. Even in situations where a full set of information is available prior to the commencement of the work, there are inevitably situations that arise which require the production and exchange of new information. This is especially so in the case of refurbishment and adaptation projects where the exact extent of the work is often not known until the building is 'opened' up. The site progress meetings provide a forum where the managers and designers can interact in a way that has similarities to a decentralised network. Decentralised structures have been found to be more effective when dealing with complex problems and also help reduce information overload (Kreps 1989).

## Construction meetings

For certain types of interaction, such as complex negotiation, Hastings (1998) suggests there is no substitute for face-to-face meetings. In a decentralised structure, such as a meeting, members have more avenues of interaction and may be able to seek assistance from someone considered supportive. The findings presented here suggest that when construction teams engage in problem solving that involve large amounts of complex information, a decentralised environment, where participants can openly engage with each other, would offer greater potential to resolve problems. Management and design team meetings provide a vehicle for information to flow freely. Further investigation is required to determine whether the contribution by specialists is unobstructed and to what extent members' interaction is influenced by others.

Discussions between management and design professionals are essential if individual aspects of the building are to be successfully integrated. Similarly, interaction between the specialists is necessary in order to determine whether the individual parts and products will function and fit together. Meetings play an important role. A number of different types of meetings routinely take place during the design and construction process. These include the formal project team meetings, sub-contractor meetings, project initiation (start-up meetings) and hand-over meetings as well as the more informal meetings arranged to discuss a specific issue (e.g. team building meetings). Common to every construction contract, regardless of procurement route, is the progress meeting. Key participants are brought together here on a regular basis, therefore it provides an ideal forum to research interaction between professionals.

Construction progress meetings are fundamental to the smooth running of the construction contract. The site meeting serves as a forum to discuss the technical co-ordination of the work as well as helping to develop and maintain relationships between the actors who have the most influence and control over the project. The purpose of meetings as a forum to resolve misunderstandings and reduce friction, while allowing discussion and decisions to be made to allow the work to proceed, seems to have changed little since it was noted in the Simon report (1944). Decision-making is an essential part of construction allowing the process to move forward. Problem solving within progress meetings falls within the technical and operational decision categories described by Loosemore (1992): technical decisions are said to be very important as they will affect the product and because operational decisions will impact on the process. Thus, the nature of professional interaction during meetings has the potential to affect the product and the delivery process.

Even though early studies identified the importance of construction meetings and benefits that would be gained from understanding the forum (Higgin and Jessop 1965), research on group meetings during the construction phase is scarce. Kreiner (1976), Loosemore (1996a), Hugill (2001) and Gorse (2002) provide research based on observations of real meetings in their natural setting.

There are no regulations or rules on the number of parties or organisations that should be contracted to design and manage the construction project. Equally, there are few construction publications that give real guidance on which and how many professionals should be present in meetings. General management text suggests that for economic reasons (Franks 1991) and optimum results (Calvert *et al.* 1995) the number of people attending meetings should be kept to a minimum. Early communication scholars such as Thelen (1949) suggest that the group should be just large enough to include individuals with all the relevant skills for the problem solution, this is known as 'the principle of least group size' (Hare 1976). It has been suggested that the optimum size for problem solving groups is five (Gray and Hughes 2001) and discussion groups is six (Bales 1958; Hackman and Vidmar 1970). When groups are below five, members often complain that groups are too small, although the opportunity increases for individual talk time. Similarly, members become dissatisfied with discussions when groups are much larger (Hare 1976). In construction projects it is common for the group members to number around ten, or more, which is less than ideal.

Construction professionals enter and leave the process at different times and contribute to different extents depending on their particular expertise (Emmitt 1999) and so it is likely that different professionals will attend sequential construction meetings. Kreiner (1976) found stable membership of site meetings in Denmark in the 1970s, but our experience of UK construction sites and preliminary research work prior to the commencement of this project suggested that membership was anything but stable. For example, when sub-contractors or consultants engage in activities during the construction phase they may be asked to attend meetings, and as their involvement in construction tasks reduces, their need to attend meetings may reduce. Problems with group development may be experienced when membership is not stable. Borgatta and Bales' (1953) study of group interaction using interaction process analysis (IPA) method found that where people have taken part in a series of meetings on related subjects and different people are present in each of the previous meetings, the group participation is the same as if the group had met for the first time. Bales (1950) claimed that this phenomenon was due to the fact that groups needed to go through socio-emotional development. As new members enter the group they cautiously interact, testing and checking behaviour and responses, understanding the roles and behaviour of others, and then establishing their position within the group and adopting the group norms. While an individual may be able to influence the group, their influence is dependent on the nature of the group that they have entered and the context that the group has set. Wallace (1987) believes this situation often occurs in construction projects and the associated meetings. He suggests that, although an architect and a quantity surveyor may have attended many meetings with different clients, it does not influence their participation at a given meeting with a new client or other construction professionals, so long as they have not worked together in the past. With this in mind, and considering the temporary multi-organisational nature of the construction team, it is difficult to perceive a situation where a team on one project would continue to work together on another project without any personnel changes. Equally, depending on demand, professionals may or may not choose to attend meetings. Such instability may affect group development, and should be considered when investigating group interaction.

Although the structure of the meeting is most closely associated with the decentralised open information network, there is the possibility that, in the group context, communication may not flow freely. Within organisation settings, some people are less willing to communicate ideas than are others (Daly *et al.* 1997). The group's norms may also influence individual members to accommodate desired communication behaviour (Kreiner 1976; Wallace 1987; Loosemore 1996a). Some members may have a greater influence on the group interaction and decision-making. This could change what is perceived to be a free flowing network into a network that does not necessarily have open lines of communication. Factors such as communication dominance, influence and reluctant communicators may affect the group's interaction. Such attributes are likely to be important in the decision-making process that leads to success or failure of the project.

### *Directing meetings*

Some interesting findings have emerged regarding participation and the person who influences discussion. Traditionally, architects have been perceived to be the leader of the construction project in the United Kingdom (Wallace 1987; Gameson 1992; Emmitt 1999). It is suggested that the external and self-perceptions of this role (as architect and historical team leader), by default, allow them a certain level of participation (Lawson 1970, 1972; Wallace 1987). Using this status he or she may assume the role of chairperson. In non-traditional procurement environments, where the contractor or other professional is perceived to be the leader, it would be expected that they assume similar participation rights (acting as chairperson and leader), as described in Kreiner's (1976) thesis. Wallace (1987) found that assumed participation rights continue until the group's regulatory forces impose a change in the situation. As this study is taken from the contractor's perspective it is important to identify how the representatives are using communication to achieve the contractor's desired results.

Boyd and Pierce (2001) interviewed project managers to find out what they actually do in practice. They found that project managers observed people to see how they were reacting to a particular situation and adjusted their behaviour in order to get the best out of the other participants at the meeting. The project managers claimed to understand the people they were dealing with, for example knowing who responds to gentle persuasion and who responds to a more aggressive stance. This research helps to illustrate how individuals use communicative behaviour to influence and react to others during meeting. While such observations are helpful, it is also important to know the communication behaviour that has the greatest effect and influence. When dealing with complex technical problems that involve multiple parties, interaction must develop the task and social requirements of the solution.

## Misunderstandings and conflict: The nature of interaction

Many of the communication problems seem to occur during the construction phase when the architect and contractor have to work closely together: a period during which the level of information, and the pressure to perform well, is at its highest. A study conducted in a non-construction communication field found that increases in the amount of information and greater involvement of actors produces increased levels of conflict (Huseman *et al.* 1977). This would suggest that the construction phase, where the majority of professionals associated with the process are involved and where the cumulative sum of information reaches its peak, would be prone to conflict. Indeed Melvin (1979) suggests that most conflicts and problems manifest themselves during the construction phase. Thus, research is required to uncover the attributes of those participants who can manage conflict and maintain relationships.

There is little information on the nature and type of interaction that takes place during site progress meetings. The few studies that do exist reveal some important characteristics relating to interaction practices. Architects have been found to dominate discussions during the initial stages of the pre-contract design phase when architectural development is needed, but reduce their contribution when other issues become important and other professionals impose their views more assertively (Wallace 1987). Communication studies have found that the interaction content of architects has tended to concentrate on architectural and design related issues and quantity surveyors tend to increase participation when aspects of cost are discussed (Derbyshire 1972; Wallace 1987; Gameson 1992). Group members are not always receptive to new information: if a view is to be considered and synthesised into the group decision it may have to be delivered assertively and other proposals may need to be challenged. During pre-contract meetings with potential building clients, Gameson (1992) found that communication between the contractor and client tended to concentrate on certain procurement options, project time-scale and services. Derbyshire (1972) discovered that contractors tended to become more involved in discussions when aspects of management are discussed.

During meetings with other actors, engineers have been found to become more involved in discussions that have a structural emphasis (Derbyshire 1972). Gameson (1992) found that during pre-contract meetings with clients, discussion between consulting engineers (structural and mechanical) and clients tended to concentrate on issues of structure, foundations, services and costs. Swaffield's (1998) doctoral research on mechanical and electrical (M&E) cost advice found that poor communication between clients and construction professionals led to misinterpretation of mechanical and electrical requirements.

Gameson (1992) also noted that differences emerged in the content of interaction when actors were engaged in discussions with clients: the professionals concentrated on issues more related to their profession. Not unexpectedly, each professional concentrates on their own area of subject specialism, and seems to devote little attention to understanding other aspects of work. Communication practices of this nature present difficulties when attempting to integrate work packages. Each specialist must have an understanding of how their work package is affected by, and affects, other works. As actors realise that components, ideas and beliefs do not integrate conflict inevitably emerges. Dealing with conflicting ideas is an essential component of effective integration.

Gardiner and Simmon's (1992) structured interviews with construction clients found that personal differences between architects and construction managers may result in, or conversely prevent, conflicts. The different backgrounds, education and training may lead to different perceptions of what is of greatest importance to the project at any point, and this could result in disputes between the professions. Such conflicts of interest between different professionals do emerge (Loosemore 1996a). When faced with a problem that requires a multidisciplinary input, two problems emerge: the professionals will concentrate on the detail associated with their specialism (Gameson 1992) and when proposing solutions the actors will attempt to reduce their organisation's resource costs (Loosemore 1996a). Professionals tend to use interaction to influence the discussions so that the resulting decisions favour the individual and their organisation (Gameson 1992; Loosemore 1996a).

Once differences emerge between actors there is a tendency for conflict to propagate. Both Wallace (1987) and Loosemore (1996a) found that when the architect and client argued there was a good chance that the same issue would emerge in later discussions. Participatory behaviour of professionals is linked to current co-operation or conflict (Wallace 1987). To demonstrate this further, Wallace cited construction-related studies that support this notion (Mackinder 1980; Mackinder and Marvin 1982). As the construction project develops, the number of parties and organisations involved will increase. It is suggested that, as the number of parties involved in construction activities increase, feedback increases, which produces conflicting objectives and value judgements (Derbyshire 1972; Mackinder and Marvin 1982). Initial conflict tends to lead to subsequent conflict inducing conflict cycles that become self-propagating (Higgin and Jessop 1965; Yeomans 1970). The level of conflict may affect project performance (Preece and Tarawneh 1997). During the construction process, problems, both large and small, may emerge which will affect individuals and their organisations' ability to perform the tasks with the resources originally allocated to the project. Meetings, negotiations and discussions are held to resolve problems. Loosemore (1994) identified two factors associated with problem solving in construction, which could lead to a defensive attitude:

- All problems involve a redistribution of resources (possibly meaning that some will benefit and some will not).
- Solutions to problems require something to change, and the act of change is not attractive to everyone.

Conflict may emerge in construction projects as organisations defend their allocation of resources (Loosemore 1994). Gardiner and Simmons' (1992) research found that conflict can be beneficial, leading to improved performance. However, if the level of conflict threatens co-operation, it is much less desirable. The temporary construction organisation is subject to communication problems. The complexity of the construction industry and ineffective communication increases the potential for dysfunctional conflict (Emmitt and Gorse 2003). When conflict develops, actors must ensure that interaction is managed so that the relationship is not permanently damaged.

### Misunderstanding and reluctance to ask questions

Misunderstandings during construction have been identified as a major contribution to legal disputes (Lavers 1992; Needham 1998). Inexperienced clients have little knowledge of the construction process and rely on the advice offered by construction professionals (Gameson 1992). Occasionally, participants may not have a full understanding of a specific construction component, activity or service. Investigations by Lavers (1992) and Lee (1997) found that professionals in this situation may be reluctant to ask for advice, or fail to inform the person relying on their services that they are unable to carry out part of their duty. Thus, professionals may offer advice or provide a service while not fully understanding the situation, even though there is a legal obligation to disclose that such advice is based on limited, possibly flawed, knowledge.

Professionals may not admit that they do not understand a situation because they fear that others will perceive them to be incompetent (Lee 1997). Although professionals should be proactive in stating their limitations, advising of the need for further consultancy services, they may be reluctant to do so. Lavers (1992) claimed that clarification constitutes good practice and is part of a legal obligation of the construction professional. Legal disputes arising from building failure often derive from a mismatch of knowledge and expectations. Through a review of construction case law, Lavers notes that it is not uncommon for a mismatch in understanding of construction or design knowledge to occur. Lavers (1992) suggests that where information is required, is missing or not fully understood, requests for information should be made to reduce potential project failure. Pietroforte (1997) identified a need for informal and responsive communication to ensure the success of construction contracts. However, the extent that such interaction occurs requires further investigation. For actors to communicate effectively, they must have an inclination of what the other person might understand, and make assumptions about their knowledge and experience. Brownell *et al.* (1997) made this observation when addressing the generic issue of communication. Assumptions about what a professional may or may not understand may be based on knowledge of the other person's profession and status (Gameson 1992). Once the speaker is aware that the addressee has an understanding of an area, he or she may be able to discuss that subject (Sperber and Wilson 1986). If a certain amount of knowledge and understanding regarding a situation is not possessed by the parties communicating, issues must be discussed and congruent understanding developed (Brownell *et al.* 1997). Macmillan's (2001) review of team work claimed that those members who are prepared to explain their assumptions and decisions would work better in teams than those who are not.

From studies of interaction behaviour Bales (1950, 1970) found that people who share a considerable amount of time together, having a close relationship, are able to provide information without offering as much explanation as would normally be required. As construction teams are temporary, generally the construction professionals will only know each other for the duration of project. Professionals must quickly learn how to interact with fellow professionals and how to gain the information and facts necessary to perform in this environment. Previous research into group interaction has found that some groups give greater amounts of explanation and ask more questions than others do, helping develop understanding (Bales 1950; Gameson 1992). When communicating across disciplines, that is across professional and cultural boundaries, there is a need for explanation and providing background information so that others, not familiar with topics, can better understand issues. Where construction participants do not understand problems there is also a need to ask for information and explanation. Project time constraints mean that the period available to understand and act on information is restricted. How professionals use interaction to overcome such barriers is clearly an important area. Further investigation is necessary to determine whether higher or lower amounts of information, explanation and question asking, during group interaction, is related to certain project success criteria.

# 3   Group interaction research

Communication between design management and construction management teams is a function of group interaction. Without an organised social system, individuals are limited to the tools and efforts they, and they alone, produce (Weick 1969). The accomplishment of projects is achieved though the interlocked and co-ordinated activities of people (Kreps 1989), and the larger the project the greater the number of people involved. With the contribution of others there is an increase in the number of perspectives, values and depth of expertise, usually accompanied by an increased amount of information from which decisions can be taken (Littlepage and Silbiger 1992). Effectiveness of groups and the degree of co-operation between their members will depend on the nature of the communication strategies employed (Ackoff 1966; Hollingshead 1996) and the training provided (Gutzmer and Hill 1973). According to Hare (1976) one of the clearest findings in the literature on small group behaviour is that the productivity of the group, no matter what the task, will improve if training is provided. However, Morgan and Bowers (1995) have argued that training must be suited to the context otherwise training programmes may have a detrimental effect on group performance. The lack of scientific research on construction communication makes it challenging to implement effective training programmes. Improvement of generic communication skills may be worthwhile, although training on more specific communication skills in construction teams is, in our opinion, not a realistic option until we all have a better understanding of interaction practices. Further research into the composition of construction teams, their behaviour over the life of projects and the nature of interpersonal communication practices must be undertaken in live projects to see how different communication patterns affect group, and hence project and organisational, performance. Such information is essential for identifying what constitutes good and bad practice in a construction context. Similarly, it is also important to understand how group behaviour can be moderated and changed to improve performance.

Only a few research projects on group behaviour have attempted to change and influence the nature of interaction. Weber's (1971) research on team building, using Bales' IPA method to study the effects of video feedback on nine different groups' interaction, found that simply exposing the groups to video footage of their earlier meetings moderated subsequent behaviour. Over a series of five meetings the control groups that did not receive feedback during the experiment increasingly engaged in less co-operative behaviour and became more negative. Members who watched the footage of previous interaction reduced their communication dominance and encouraged others to participate in the group. Gorse and Whitehead (2002) also found that group members became self-critical and identified behaviours that they would like to change and improve after watching video feedback of their earlier performances. So it would seem that even simple reviews of behaviour can result in changes to group interaction.

Research into group communication has shown that the process is difficult to understand in its entirety (Poole and Hirokawa 1996). The interactions of multiple actors who are subject to psychological, social and contextual influences make the subject difficult to piece together as a complete picture. With the vast number of variables constantly changing and with interaction occurring in different time frames, it can be very difficult to trace causal analysis, although interaction variables associated with successful projects can be identified. Thus, much of the research on group interaction and communication focuses on group satisfaction. These studies frequently deal with perceptions of how members feel after a particular group encounter or how they think they would feel under particular circumstances or situations. They provide a useful insight into individuals' perception of behaviour, but the work has not measured behaviour or performance. Indeed, perceptions of how people think they behave and how they actually behave can be quite different. This means that research based solely on perceptions should always be checked against actual behaviour before assuming that changes based on the research results will result in improved performance. Similarly, research taken from different contexts should always be carefully analysed to see whether the methodology and/or results can be applied to a new context, and if so, with what caveats.

## Multidisciplinary groups and organisational communication

Multigroup communication occurs within social systems that are made up of interdependent groups that share the performance of tasks to achieve a common goal (Kreps 1989). Construction teams are composed of groups of individuals with different specialist knowledge from different organisations, forming a loosely coupled multidisciplinary group. This helps to ensure that the various aspects of a project are developed in a manner befitting the client's aspirations. As a general observation the literature on group performance (Hare 1976) and multidisciplinary teams (Ysseldyke *et al.* 1982) tends to suggest that decisions made by groups are more workable and accurate than those made by an individual. Due to their broad range of specialisms, multidisciplinary teams have been found to consider a wide range of potential solutions (Ysseldyke *et al.* 1982) and groups have also been found to make more accurate, workable and rational decisions (Stroop 1932). Stroop has argued that the grouping of knowledge and experience acts as a moderating influence to restrict extreme views. The group's regulatory forces (imposed using a combination of conflict and group norms) control unacceptable views that are presented to it (Wallace 1987). Unacceptable views are moderated through reactions, argument and conflict (Stroop 1932). Contrary to these findings, Rim's (1963, 1964a,b, 1965, 1966) research on group and individual risk taking found that group decisions were more risky than those of individuals. Using Wallach and Kogan's (1959) risk-taking measures to examine the behaviour of sixteen groups of five students, Rim found that thirteen of the groups adopted higher risk strategies to problem solving following group discussions. The subjects' level of risk taking was based on the lowest probability of success they would accept when engaging in a task.

Group interaction changes the behaviour of individuals. As a general rule individuals will accommodate greater risks within a group environment than they would on their own. Using the same research approach as Wallach and Kogan (1959), Bemm *et al.* (1970) found that individuals within groups would take greater risks even if the consequences of the risk taking would affect them personally. However, when group members were informed that failures associated with risk taking would be openly discussed and disclosed to the group, there was a shift to lower-risk decision-making. The fear of group or public humiliation appears to temper risk-taking behaviour. Although such findings have been compared to the 'real world' context, the findings remain limited to laboratory-type experiments. In real life it can be very difficult to identify the level of risk associated with a decision. Similarly, it can be a challenge to state with any certainty whether the distinction made between individuals and groups applies to all commercial decisions. Organisational studies that may provide a direct comparison to our research are scarce, although reports are available on group behaviour associated with unsuccessful organisational outcomes and these are reviewed later.

Early research by Stroop (1932) found that group interaction produces a higher degree of creativity in relation to a potential solution to a problem than an individual does, although others (Taylor *et al.* 1958; Lamm and Trommsdorff 1973) subsequently noted exceptions to this observation. Lamm and Trommsdorff's (1973) and Taylor *et al.*'s (1958) research on idea generation through brainstorming exercises has shown that individuals outperform the group by a factor of 2:1, and the individuals' ideas were found to be more creative than those of the group. The main finding from their research was that the process of group pressures inhibit participation by some members. Individuals were found to participate less in small groups when they perceived their skills to be inferior to other group members (Collaros and Anderson 1969), a finding supported by Emmitt and Gorse (2003). In a parallel research project undertaken over a five-year period we found that groups of four, five and six project management students consistently underperformed in brainstorming (also known as *information showering*) activities compared with the combined results of individuals working separately. We found that turn taking, turn blocking, fear of embarrassment and the loss of concentration all had a detrimental effect on the group's ability to generate ideas. As a general finding each individual would produce fewer ideas than the group, but the combined output of the individuals outnumbered that of the *bona fide* group. Although the individuals were more productive than the groups they could not benefit from the evaluation of ideas within the group setting, which may be a disadvantage. Although some individuals find it difficult to interrupt others, gain the floor or generally participate in group discussions, when rules and structural mechanism were introduced to ensure that members take turns or all participate in the discussions the natural flow and development of decisions became further hindered, when compared to the unstructured *bona fide* group. While mechanism to ensure each member participates often reduces the effectiveness of the group, individuals who are not confident speakers often benefit from the experience, gaining confidence and improving communication skills. Over time, groups that have structures or rules imposed do improve their performance as they become more accustomed to the framework in which they have to work.

When working independently some individuals produced a comparable number of ideas to that of the *bona fide* group. However, when individuals who show signs of exceptional performance are placed together in a group, similar productivity benefits may not be achieved. Belbin (1981, 1993, 2000) found that groups of highly intelligent individuals often performed worse than that of a randomly formed group. This is because groups of highly intelligent people often have difficulty in agreeing on whose idea is the best.

Brown's (2000) review of group communication research indicates that the evaluation of ideas may be better dealt with in groups where different perspectives can be used to analyse ideas. While individuals and groups can both improve and limit idea generation and the evaluation of ideas, it is always important to consider the task and the attributes required from the individual or group. From the research and literature that surrounds idea generation and evaluation it is clear that for complex problems elements of the task should be broken up into individual and group exercises combining the attributes of individual and group problem solving.

Campbell (1968) reported that the contribution of more than one person increases the potential to solve complex problems quickly. Typically, in laboratory studies of human problem solving, subjects are presented with just sufficient data to solve the problem. However, a prime obstacle to solving 'real life' problems is selecting the relevant data from the body of superfluous, irrelevant and possibly misleading data. In line with this logic, Campbell conducted an experiment testing the subjects' ability to determine a sequence of codes from various words. Students were tested on their ability to solve the code when the data was just sufficient and where data irrelevant to the solution was included. Campbell found that the subjects generally take longer to solve problems where they had to differentiate between irrelevant and relevant data; this is something that is worth remembering when attempting to understand real life problems. More interesting, the results show that individual members could at the outset adopt a route that would lead to the solution. Some individuals appear to have the ability to instantly separate the relevant from the irrelevant data, although when more than one person is working on the same problem the chance of finding appropriate information in a shorter time period is increased. Although a laboratory experiment (and not really a group exercise), the research does provide an example of a situation where the ability to solve difficult problems is increased when more than one individual contributes. In commercial environments participants must explore the various options and solutions available to them when solving problems. Although Campbell attempted to move laboratory experiments closer to the real world by requiring subjects to select the relevant information from the irrelevant, in complex problems people are rarely limited to just one solution. Laboratory research provides insight into problem solving and decision-making behaviour; however, it is likely that group behaviour in commercial organisations is more complex.

Interaction and compatibility of group goals and individuals' goals in multidisciplinary teams is fundamentally different from the corresponding process in unidisciplinary groups (Cartwright and Zander 1968; Lieberman *et al.* 1969). In unidisciplinary groups the objectives of each individual are likely to be similar to those of other members, while in multidisciplinary groups there is likely to be larger variation in objectives (Wallace 1987). Differences between group members may make the establishment of team goals difficult and reconciling individual objectives with the overall group objective may be equally problematic, resulting in ambiguity and group conflict. Bales (1953, 1958, 1970) suggested that role ambiguity may lead to apathy towards overall goals, disrupting the early stages of group development and interfering with the implementation of a compatible social structure for the group. Establishing the aim of the group and the role of the individual and developing a group structure that aligns individual skills towards the goal are important. Incompatibilities and ambiguity will cause problems and so conflict within the group may be a natural way of (re)aligning group behaviour or conversely disrupting the group as it develops, changes and new information is considered by group members.

Multidisciplinary teams have also been found to propose and consider a wider range of solutions to a problem when attempting to arrive at an overall solution (Ysseldyke *et al.* 1982). Although Bales suggests that multidisciplinary teams may appear more productive in terms of alternative solutions generated during interaction, this could be a result of goal ambiguity. Yoshida *et al.* (1978) examined the content of multidisciplinary group interaction and classified it into: contributing information, processing information, proposing alternatives, evaluating alternatives and finalising decisions. They found that the frequency of the individuals' participation, their perceptions and their contribution to multidisciplinary teams varied more than that in unidisciplinary teams.

Yoshida *et al.* found that the stronger combined group forces often overruled individual expertise and experience. Thus, group consensus may go against expert opinion and information. Contrary, work by Littlepage and Silbiger (1992) found that, regardless of uneven and skewed participation rates, groups were able to recognise and use individual expertise confidently. In their study, 324 college students were assigned to 1, 2, 3 or 10 person work units and were asked to complete 20 multiple choice questions. When answering the general knowledge questions they were also asked to indicate how confident they were that the answer given was correct. A second part to the study controlled participation in the 10-person groups. Subjects were given 5 coloured cards and 10 white cards. Members deposited one card in a box each time they spoke. The coloured cards were deposited first. Once a subject had deposited all of their cards he or she was not allowed to speak again until all other members had deposited all of their coloured cards. The use of the cards resulted in a greater distribution of participation amongst members, but it did not increase the groups' ability to recognise individual expertise. Participation may not necessarily be the most dominant influence in recognising group expertise.

When repeating this experiment we found that the use of the coloured card system resulted in 'stuttered' interaction. Those most reluctant to contribute largely controlled the pattern of interaction. Reluctant members still suffered from an apparent inability to contribute at the same speed and with the same fluency of other members. When the exercise was repeated a number of times, the most confident interactors improved their ability to encourage the more reluctant members to contribute. Eventually the flow of the participation improved and the reluctant members' contribution was strengthened. Over time, the learning gleaned from repeated exercises helped to improve the group participation dynamics. These experiments seem to suggest that group members need to practice to make more distributed patterns of interaction work. To do this they need to be educated and trained in the use of turn taking and helping others to communicate.

Gameson (1992) found that construction specialists would rely on their own knowledge rather than suggesting that others should be consulted. Specialists were more inclined to sell their specialism rather then recognising the attributes and knowledge of other specialists. The competitive nature of professionals does seem to have an effect on professionals, such that they avoid exposing their knowledge gaps to others. Lee's (1997) investigation, based on a laboratory experiment involving 153 participants and a hospital field study, made similar observations; results showed that even when professionals experience difficulties in solving a problem due to limitations in their knowledge, they tended to rely on their incomplete knowledge rather than consulting a specialist for help. Lee found that this was particularly prominent in male participants, although as the females assumed higher status positions they also became associated with the trait of not openly asking for assistance. Mabry's (1985) laboratory study of 44 student groups using the Bales' IPA method found no difference in the level of questions used by groups composed predominantly of males or females. So, gender difference may not be an issue in the use of question asking and information seeking, although construction participants who think they hold key positions may be reluctant to seek help, request more information or ask questions that may be perceived as revealing gaps in their knowledge. High status professionals may still use probing questions to contest the knowledge of others, query the nature of information and/or to undermine other contributors, but may avoid openly asking for help.

### *Effective communication*

Communication is critical to business. The information fed into and subsequently held within the organisation is used to make decisions (Cyert and March

1963; Kreps 1989). Effective communication in decision-making is required to share understanding of problems and discuss the various solutions (Gouran and Hirokawa 1996). Hosking and Haslam's (1997) observations of business relationships found that informal conversations within organisations were an important process for understanding what were considered as 'taken-for-granted' statements; thus, conversation was essential to overcome ambiguity. The importance of background information, clues and what may be considered 'small talk' is important for building relationships and thus crucial for developing an understanding of unfamiliar contexts. Being able to enquire further into subject matter without the fear of embarrassment, ridicule or risk of offending others is achieved primarily through interpersonal interaction, which helps to build relationships and hence establish contextual information. Research is required to show how such behaviour is structured and takes place in a construction context.

Hollingshead (1998) identified the importance of overt interaction to decide responsibility. Her laboratory experiment, which involved 88 couples tasked with learning and later recalling words in 6 different knowledge categories, found that different communication practices affected the ability of the dyad to recall information. Hollingshead found that when members of a group are tasked with a problem, they became specialists in some areas but not others, and all members came to expect each participant to access information in specific domains; thus, each individual took responsibility for specific tasks. Specialisation reduces the cognitive load on the individual, while also providing the group with access to larger amounts of information. Newcomers must communicate to explicitly identify responsibility for gathering and processing specific information; making assumptions about responsibility for problem solving results in less effective teamwork and duplication of tasks. In groups it is necessary to know who is most knowledgeable and skilled in specific areas so that people can assume key roles in related tasks. When members freely interact and openly disclose information other members gain access to, and clues about, a member's knowledge and skills. Such information is key to establishing informal group roles and hence the effective use of the group's knowledge (Figure 3.1).

Over time, the knowledge of each member's skills and attributes should make the group more effective. Roles and responsibilities can be assumed and the most appropriate person can be quickly allowed to undertake the task, without the need for lengthy discussions to determine who has the necessary skills or knowledge. Effective groups in industrial settings are those that are more productive and meet the organisation's objective. Shepherd (1964), drawing on previous communication research, suggests that successful groups have open and full communication; information, ideas and feelings are exchanged, and, no one holds back. However, without placing this view in context with other communication studies, the statement is quite naive.

A good example of interaction where feelings are exchanged openly and 'no one holds back' can be seen in the behaviour of children, which Bales (1950) described as unorganised and unrestrained. The behaviour of adults is far more sophisticated, with emotional interaction often restrained to a bare minimum

Exchanges are needed to establish and structure the relationship. As information is exchanged, important attributes of each individual will be shared and used.

Exchanges help to disclose:

Skills
Knowledge
Expertise
Experience

and

Willingness to share and help

*Figure 3.1* Identifying individual attributes and using them within a group situation

(Bales 1950). Interaction differences in adult groups are often subtle (e.g. Bales' 1970 review of 21 adult group studies). Thus, while restrained by adult norms, more open exchanges of ideas and emotions, as described by Shepherd, may be more productive. Overuse of emotion can easily stray outside what may be considered 'normal' behaviour, yet relevant use of emotion is known to be highly effective (Goleman 1996; LeDoux 1998). The use of appropriate emotion can be used to build relationships and help to generate desired behaviour (Goleman 1996). While adult behaviour tends to restrain use of emotion, being able to respond to and use emotion is important for building, maintaining and structuring effective groups.

Groups that have been found to be more productive are said to have a structure that is suited to their function (Hare 1976). The high level of productivity is achieved not only because they have procedures for solving problems but because the group is stable and less time is devoted to status struggles (Heinicke and Bales 1953; Hare 1976). Similarly, the members are aware of each other's skills, attributes, knowledge and roles and only a relatively small amount of discussion is required to organise tasks.

Various arguments have been presented regarding what improves the effectiveness of group communication; however, the nature of the improved interaction is often task specific. Research by Meister and Reinsch (1978) and Woodcock (1979) found significant communication skill deficiencies often arise at times when effective communication skills were most needed.

## Group development and group norms

The development of the construction team will have an effect on the group's interaction and vice versa. Similarly, the strength of the professional relationship between designers and managers and their ability to resolve problems will also be a function of their interaction behaviour. Achievement of a group's goals depends on concerted action and so group members must reach some degree of consensus on acceptable task and socio-emotional behaviour before they can act together (Hare 1976). Interaction has been described as task specific and social. The social element of interaction is developed through emotional exchanges that are used to express a level of commitment to the task and other members. To accomplish group tasks, relationships need to be developed and maintained. The level of interaction associated with maintaining, building, threatening and breaking-up relationships will be a function of socio-emotional interaction and will be subject to group norms.

A group's behaviour develops and changes over the period of interaction. Even the behaviour of children becomes more structured over a series of meetings (Socha and Socha 1994). As task groups attempt to solve problems, they undergo changes in terms of their attitude and behaviour towards each other (Hoffman and Arsenian 1965). Groups go through a process of learning (Wallace 1987), which can result in changes to structure as the group moves through a range of social, emotional and developmental stages (Bales 1950). Two variables that are said to affect the group development are the length of time that a group has existed (Lieberman *et al.* 1969) and the number of occasions that the group has previously met (Hoffman and Arsenian 1965). Both factors are important in a construction context. Borgatta and Bales (1953) studied groups using the IPA method and found that when people have taken part in a series of meetings on related subjects and different people are present in each of the previous meetings (a common occurrence in construction meetings), group participation is the same as if the group had met for the first time. This would tend to suggest that construction meetings are not overly productive. Bales (1950) found that this phenomenon is due to the group's socio-emotional development. Individuals are not aware of the group's social and emotional norms, nor does the group know how the individuals will react to the group norms. Thus, a socio-emotional framework develops and re-establishes itself when new actors enter the group.

Initial group meetings necessitate the formation of socio-emotional structure and a participation strategy that will be used to access information, develop group knowledge and make decisions (Cartwright and Zander 1968; Shadish 1981). When actors have experience of the subject being discussed but have not experienced interaction in that group setting, they are more likely to use task-based interaction rather than socio-emotional interaction (Bales 1953). Socio-emotional communication emerges later once the newcomers are familiar with the group members and the group norms. Although task-based discussions build understanding, socio-emotional interaction is a fundamental part of any decision-making process. Agreeing and disagreeing are socio-emotional communication acts (Bales 1950). Equally, emotion is used to show support and concern. In groups it is the socio-emotional interaction that regulates interaction and provides the framework against which decisions are made.

## *Development of group norms*

Although the behaviour and characteristics of groups change and develop over time, it is well known that groups develop (and are subject to) behavioural norms. Newly formed groups will develop relatively stable patterns of interaction as they mature (Heinicke and Bales 1953; Keyton 1999). Scheerhorn and Geist (1997) describe group norms as the recurrent patterns of thinking and behaviour. Anderson *et al.* (1999) make a distinction between rules and norms, noting that members come to accept norms as their way of being a group and doing group work, whereas rules are agreements about how to behave appropriately. The norms of group behaviour may be specifically associated with the reason or purpose the group formed, or the task, or they may be attributable to the group make-up.

Group norms may change as members adapt to changes in context. In almost every situation there are a number of specified roles, environmental clues or repertoire of acts that provide information about how the individual is supposed to interact (Jackson 1965), and these vary from one situation to another (Furnham 1986). It is important to note that the norms found in laboratory groups (groups created for the purpose of studying group behaviour) have been found to be different from real life groups, or *bona fide* groups (Ketrow 1999).

The expectations of the way members are supposed to act are articulated into implicit rules that are adopted by the group to regulate its members' behaviour (Jackson 1965; Fledman 1984). Such norms and rules are said to provide powerful control over the group. While there are rules and norms that are explicit, it is those that are implicit that have the greatest direct effect on relational behaviour (Keyton 1999). It would seem that norms are the least visible yet most powerful form of social control that can be exerted on a group (Bettenhausen and Murnighan 1985).

Group norms can be so influential that some members will express a judgement differing from the one they hold privately (Hare 1976; Hackman 1992; Schultz 1999). Fledman (1984) has identified four ways that norms are developed:

- Norms can develop from behaviour and statements made by leaders and other influential members.
- Critical events in the group's history can establish a precedent. Keyton (1999) offers an example of this, suggesting that, when members are faced with a deadline, the group may change from the normal leisurely pace of interaction to a faster pace to ensure the deadline is met. The group uses previous responses resulting from tasks and demands undertaken that have set an implicit response or framework for the situations and other similar situations.
- Norms may simply develop from repetitive behaviour patterns. Such patterns are particularly prominent in certain seating configurations (Ketrow 1999), formal events and meeting procedures (Sunwolf and Seibold 1999).

- Members can import group norms from previous group experiences (Katz and Kahn 1978; Anderson *et al.* 1999). Where members have worked in other groups they may assume that a new group operates in the same manner, and if enough members accommodate the behaviour it will develop into a group norm.

Newcomers need to observe the communication behaviour and practices of other members so that they can understand the group culture and participate in it (Trujillo 1986). When new groups form, they establish beliefs, values, norms, roles and assumptions as a result of communicative behaviours that uncover similarities and differences (Fledman 1981; Anderson *et al.* 1999). An individual's actions and behaviours are also influenced by his or her motives for membership, positions and role (Zahrly and Tosi 1989). It is clear that norms formed through communication affect group interaction and decision-making.

### Norms and decision-making

A number of case studies focus on how norms are used by groups (Hirokawa and Salazar 1999). Rules and norms are habitual, forming a backdrop or structure against which decisions are made (Larson and LaFasto 1989; Hackman 1992). These and other studies have shown that norms can have positive and negative effects on the decision-making process (Janis 1982; Senge 1990). They can encourage cohesion and agreement, suppress critical inquiry (Janis 1982; Cline 1994), reduce political input and increase rational discussion (Senge 1990).

Looking more specifically at language, Giles (1986) noted that communication behaviour reflects the norms of the situation; however, it is often the communication behaviour and language that is used to define, and subsequently redefine, the nature of the situation for the participants involved. Group norms are powerful and will affect the way actors interact within the group; however, some individuals are able to exert such a strong influence that it changes the group norm.

### Individuals who deviate, and are exempt, from norms

Group norms affect all members of the group; however, Keyton (1999) has suggested that high status members might be exempt from the norm expectations that others are expected to follow. Generally, it is assumed that if a member deviates from the group norms the other members will react in one of three ways (Hackman 1992):

- Group members may try to correct the behaviour, normally through pressure outside the group environment (a form of informal diplomacy).
- If deviation persists, other group members may exert psychological pressure through communication within the group, placing the deviant in an 'out-of-group' position (which the majority of people find uncomfortable).

- Finally, if deviation presents an acceptable alternative to the group norm and the behaviour (stance) is maintained over time it can influence other group members to accommodate the alternative norm.

The way a group interacts and behaves can be changed radically by the presence of one or two newcomers. For example, one may consider the actions of a group of children playing and the change of behaviour when the school bully enters the group, or the behaviour of teenagers and the changes that take place when parents enter their environment. The most striking change of behaviour in professional settings seems to occur at relatively informal events, for example sports functions or in corporate boxes etc. when a chief executive unexpectedly enters the room or comes into close proximity of colleagues who would not normally interact with their highest ranking professional. Informal chit-chat and banter often stops when the director or CEO arrives, the mood changes and the focus is now on the high status member. The actions and behaviour of the chief executive then seems to be the strongest indicator of the group behaviour that follows. Although anecdotal, these recollections of high status individuals changing group behaviour are easily recalled.

## Bales' interaction process analysis

Much of the research reviewed in this book so far has used content analysis to observe, record, measure and hence analyse group interaction. Most of this work has been developed from the seminal work of Bales (1950), which made a clear distinction between task-based and relational communication (Keyton 1999) and conceptualised the group as a social system (Mabry 1999). The early work of Bales (1950, 1953) did two things: it provided a system for categorising communication acts (Interaction Process Analysis), and it established the theory of phase movement during group communication (equilibrium theory). IPA provides a detailed method for observing and coding group members' communicative behaviour, at a micro-level, so that their interaction can be isolated, recorded, and subsequently interpreted (Schultz 1999). Bales' IPA is one of the most widely used research techniques to study overt group interaction in the social sciences (Mills 1967; Hartley 1997), with application to the construction field limited to the doctoral work of Wallace (1987), Gameson (1992) and Gorse (2002).

The IPA system specifically focuses on task and emotional communication acts (Figure 3.2). The task acts are categorised to capture neutral emotional exchanges of information, the exchanges needed to achieve group goals. The emotional categories are separated into negative and positive reactions. The socio-emotional categories are those used to develop and maintain group relationships (Hare 1976).

The publication of Bales' IPA shifted attention from issues of reasoning and the individual's oral style to trying to explain how patterns of communication constitute systemic processes, influencing group integration and performance (Mabry 1999). IPA provides a method of breaking communication down into

| NUMERICAL ID | CATEGORY DESCRIPTION | | |
|---|---|---|---|
| 1 | **SHOWS SOLIDARITY** – raises others' status, gives help, encourages others, reinforces (rewards) contribution, greets others in a friendly manner, uses positive social gesture. | F | **SOCIAL-EMOTIONAL AREA** Positive reactions. Behaviours used to encourage commitment, help build and strengthen relationships. |
| 2 | **SHOWS TENSION RELEASE** – jokes, laughs, shows satisfaction, relieves or attempts to remove tension, expresses enthusiasm, enjoyment, satisfaction. | E | |
| 3 | **AGREES** – shows passive acceptance, acknowledges understanding, complies, co-operates with others, expresses interest and comprehension. | D | |
| 4 | **GIVES SUGGESTION** – makes firm suggestion, provides direction or resolution, implying autonomy for others, attempts to control direction or decision. | C | **TASK AREA: NEUTRAL** Input and attempted answers. Acts used to develop information, understanding and control. |
| 5 | **GIVES OPINION** – offers opinion, evaluation, analysis; expresses a feeling or wish; seeks to analyse, explore, enquire; provides insight and reasoning. | B | |
| 6 | **GIVES ORIENTATION** – provides background or further information, repeats, clarifies, confirms; brings relevant matters into the forum, acts that assist group focus. | A | |
| 7 | **ASKS FOR ORIENTATION** – asks for further information, repetition or confirmation. Acts used to request relevant information and understand the topic. | A | **TASK AREA: NEUTRAL** Questions and requests Acts used to seek, analyse and explore information and request direction. |
| 8 | **ASKS FOR OPINION** – asks others for their opinion, evaluation, analysis, or to express how they feel. Act used to request or explore reasoning. | B | |
| 9 | **ASKS FOR SUGGESTION** – asks for suggestion, direction, possible ways of action. Requests for firm contribution, solution or closure to problem. | C | |

*Figure 3.2* Bales' 12 interaction categories (Adapted from Bales 1950: 9)

| 10 | **DISAGREES** – shows passive rejection, formality, withholds help, does not support view or opinion, fails to concur with view, rejects a point, issue or suggestion. | D | **SOCIAL-EMOTIONAL AREA** Negative reactions, Behaviours used to |
|---|---|---|---|
| 11 | **SHOWS TENSION** – shows concern, apprehension, dissatisfaction or frustration. Persons interacting are tense, on-edge, Act that express sarcasm or are condemning. | E | reject task information, question commitment and threaten relationships. |
| 12 | **SHOWS ANTAGONISM** – acts used to deflate others' status, defends or asserts self, purposely blocks another or makes a verbal attack, expressions of aggression and anger. | F | |

Examples of communication associated with each category can be found in Appendix A1

**LEGEND**

| | |
|---|---|
| **A** Problems of orientation | **D** Problems of decision |
| **B** Problems of evaluation | **E** Problems of tension management |
| **C** Problems of control | **F** Problems of integration |
| As task acts move from A to C greater effort is extended towards control and closure. | Emotional intensity of communication act increases from D to F. |

*Figure 3.2* (Continued)

predefined categories, from which the structure of group interaction can be analysed. To some extent the predefined categories determine and hence limit the data being captured; nevertheless, the method allows for scientific investigation into task and emotional interaction. It can be used to look at interaction trends, more specifically it can be used to identify where and when certain types of behaviour occurs and to identify responses to those behaviours.

## Strengths, limitations and application of Bales' IPA

Bales' IPA system and the studies that have developed from it (Bales and Strodtbeck 1951; Bales *et al.* 1951; Psathas 1960) have been particularly influential in helping to develop an understanding of interaction as a part of the group process (Gouran 1999). The work of Bales and others using the IPA method has provided a basis for the view that groups constitute systems; and that communication within the group exhibits regularities, especially in the development phases of group interaction (Gouran 1999). Mills (1967) and Brown (2000) have claimed that there are few methods for collecting empirical data of group interaction and behaviour that are better than Bales' IPA. Furthermore, they suggest that the method has provided important insights into group interaction, classifying direct, face-to-face interaction as it actually happens.

Bales' (1950) initial argument was to assume that any group is ultimately directed towards the achievement of a task. Thus, the group has a function and moves through different stages of interaction behaviour to achieve that function. Using the IPA coding system, analysis of the group and individuals' interaction is possible. The empirical method can provide operational substance to interaction, showing the stages, phases, conditions and occurrences that can result in specific conditions or problems (Bales and Strodtbeck 1951). The method has led to the discovery of certain regularities and tendencies (Talland 1955; Psathas 1960; Mills 1967; Brown 2000). Once identified, the patterns of interaction that tend to result in specific behaviour can be used to great effect, or if necessary the interaction can be moderated so that the undesirable outcome is avoided (or mitigated).

The main strength of Bales' IPA is its ability to examine different types of groups. The method offers a generic system that is not specific to one context (Stone *et al.* 1966; Olmsted 1979). The methodology has been successfully used to analyse: labour mediations (Landsberger 1955); three-person family units that contained a normal or schizophrenic child (Cheek 1965); undergraduate training groups (Stone *et al.* 1966); leadership (Katzell *et al.* 1970); construction professionals during initial meetings with building clients (Gameson 1992); issues confronting both co-located and virtual design teams (Bellamy *et al.* 2005); children's task group communication (Socha and Socha 1994); primary school children, adolescent groups, groups of chess players, university discussion groups, and interaction between married couples (Bales 1950); 21 different studies of adult groups (Bales and Hare 1965); adult group norms (Bales 1970); Bell's (2001) observation of child protection teams; and interaction in construction teams (Wallace 1987; Gameson, 1992; Gorse 2002). The Bales' IPA system is said to offer a full repertoire of verbal acts, although a few limitations have been noted (Furnham 1986).

It is evident from a review of the literature that the Bales' IPA may be successfully applied in a wide variety of situations, although a few researchers have noted some difficulties in specific situations. For example, Socha (1999) found that the system was difficult to apply to studies of the family unit. McGrath (1997) claimed that the Bales' method was formulated as a tool to discover and support a theory. McGrath claims that no matter how general a functional coding system may seem, and no matter how neutral the categories, the system always represents a particular perspective and thus may not be useful in all situations. The problem with any classification system is the categories are predefined and determine what information is collected. Data that does not fall within the categories is missed or assembled with a category perceived as closely related. Poole *et al.*'s (1999) critique of IPA research explains that special coding functions may be necessary in some 'special' situations, such as the examination of legal proceedings.

Bales' IPA has 12 categories, illustrated in Figure 3.2. Each category is developed from a vast body of explanation that is supported by rules and these are applied regardless of terminology that is specific to context. Indeed, as already discussed, the system is generic, allowing the researcher to ignore terminology and concentrate on the intended nature of task-based and socio-emotional interaction. The system is not used to evaluate the content of a specific communication act (unlike, e.g. audio recording and transcription). Bales has argued that if a specialist system for identifying subject-specific interaction is developed it is limited and tightly bound to the original research context, and this makes it difficult to use for other research. However, a context-specific system is often required. Wallace (1987) combined and developed a bespoke system, using parts of Bales' IPA and other methods; however, the combination severely limited its potential for use in other contexts. Because Wallace developed his own system with additional categories mixed in with the IPA system it is not possible to compare the work of Wallace to others who have used the IPA system in its original form. Gameson (1992) took a slightly different approach, using two systems separately, IPA and a bespoke system. The IPA system was applied in its original form in accordance with the Bales (1950) protocol. Gameson (1992) stressed the importance of using the documentation provided by Bales for producing reliable and valid data. This means that the results pertaining to Gameson's research can be compared to other research using the IPA method, whereas Wallace's research cannot. When considering how data is to be collected from within construction teams it may be necessary to use one or more systems so that data can be triangulated; however, we would urge researchers not to alter well-established and respected systems, simply because it prevents comparison with earlier work. Given that we need to build up a strong research base in construction communication the tendency to invent new systems should be tempered until we have developed a better understanding.

### Equilibrium theory

Using the IPA method Bales and Strodtbeck (1951) uncovered evidence to suggest that problem solving groups exhibited recurrent patterns of interaction, which they identified as orientation, evaluation and control. The initial phase 'orientation' is marked by high levels of task-related messages in the form of information, opinions and suggestions, and positive and negative reactions to this information. This phase involves communication about the nature of the problem to be solved, when actors orientate themselves within the group, identify aspects of the task and contemplate their role. The second phase 'evaluation' is characterised by a reduction in exchanges of informational acts, a levelling out of opinion and evaluation, accompanied by suggestive behaviour and positive and negative reactions. During this phase the group confronts 'what to do and how to do' type issues. At this stage roles and responsibilities start to be outlined and the group structure emerges. In the final stage, the group establishes 'control', this is marked by a continued and sharp decline in informational behaviour, a slight decline in opinionation, a reduction in the quantity of suggestions and negative emotional behaviour, and continual increase in positive reactions. The control phase involves deciding what to do. At the same time as the task-related acts are discussed, a parallel cycle of positive socio-emotional phases results; interaction acts such as showing solidarity and tension reduction are used. During the final stages the structure of the group has developed and is relatively stable allowing tasks to be identified, allocated or adopted. Roles within the group are also relatively clear at this stage. Bales (1953) found that in order to address problems, groups had to move through this set of acts. To ensure that task-based discussions can continue, the relationships are maintained with positive and negative emotional exchanges. The structure of the group needs to be maintained and adjusted as necessary taking into account the nature of task and reactions of individuals.

Theories such as the equilibrium theory, because of their single-dimension approach, became known as linear-type theories. This approach went largely unchallenged until the 1970s and 1980s (Gouran 1999). Scheidel and Crowell (1966) were the first to question this theory. They noted that interaction did not unfold in a linear fashion and needed to take into account other internal and external stimuli. According to Scheidel and Crowell (1966) interaction is determined by a set of internal relationships that emerge from the elements composing and characterising a group; at the same time the group is influenced by other elements from external environments. Inspired by the work of Scheidel and Crowell (1964, 1966) Poole (1981, 1983a,b) also found evidence to support the multiple sequence models. The view was that group interaction progressed in many different fashions, constantly changing and adapting, rather than through linear development. Poole (1981) argues that group acts are contingent on task and situational factors, and that multiple sequences of phase movement are caused by such variables as task characteristics, group composition and the level of conflict evoked by task issues. Bales (1950) had also claimed that the same factors affected interaction, but this appears to have been overlooked by others. The arguments over unidimensional and multidimensional methods of studying interaction continue; however, theories developed from both approaches have contributed to the understanding of group dynamics. Neither approach offers a comprehensive picture of the multidimensional nature of group communication, but both provide valuable insights into the interaction process.

### System for multiple level observation of groups (SYMLOG)

Following the development of the IPA technique, Bales and Cohen, with assistance from Williamson (1979), developed a different system for studying groups, called SYMLOG (System for the Multiple Level Observation of Groups). This method is an extension of the IPA system, and although it is considered to be theoretically more complex, it is believed to be more flexible for collecting multiple perspectives (Keyton 1999; Poole *et al.* 1999). Rather than limiting observation to an independent researcher, the system is based on participant observation of others and a self-study of internal feelings. The advantage of participant study is that observations are not just limited to overt interaction and behaviour, but they also capture the participants' own feelings, judgements and values of themselves and other individuals within the group. SYMLOG can be used to enquire into areas of group interaction that the IPA system cannot, but it requires participants to complete a number of proformas. Each proforma provides reflective data on the participant's motives, feelings and contributions as well as their perceptions of other members' behaviours and interaction attempts. Tables and grids are used to establish some consistency to the feedback gained from each participant. With the increased number of people coding interactions, there are increased difficulties experienced with intercoder reliability; and functions may be attributed to ambiguous acts (Poole *et al.* 1999), although the forms and tables do provide a degree of consistency over the subject matter that would not be achieved if each participant's reflection of events was totally open and unstructured. The time for each individual member to understand and complete the SYMLOG self-study is about three to four hours for a group of five; larger group sizes require more time (Bales 1980). Furthermore, SYMLOG is extremely complicated to apply because of the depth of multiple analysis (Poole 1999), and in live business projects there are real difficulties associated with applying this type of methodology. Considering the difficulties experienced by Loosemore (1996a, 1998b) when requiring participants to complete diary entries, the use of SYMLOG, which uses a small booklet to collect data from each participant, taking in excess of three hours to complete, would be impractical. Poole (1999) notes that while the IPA has received considerable attention, the SYMLOG system is hardly used by group communication scholars, possibly because of the difficulties of application. Fryer *et al.*'s (2004) SMOT analysis is easier to apply than the SYMLOG, yet still requires a considerable amount of time by each participant to state their motives and thoughts regarding each group meeting. Obtaining consistent feedback from professional participants following discussions and other encounters can be particularly difficult (Loosemore 1996a, 1998b; Gorse *et al.* 2000a,c).

# 4  Group participation and interaction

Group participation and interaction are key characteristics of project-orientated organisations. Participation is the extent to which individuals are involved in group interaction, being coloured by the group norms and group development (Wallace 1987; Anderson *et al.* 1999). Research into participation should also include aspects of turn taking, the initiation of conversation, how individuals interrupt others and if, and how, the interaction intensifies. Littlepage and Silbiger's (1992) research on participation and turn taking found that interaction is controlled and dominated by a few members of the group. In moderate and larger sized groups it is widely accepted that participation among group members is skewed and unevenly distributed, but this does not necessarily hinder group performance. Although participation is uneven there is some evidence to suggest that group members become more dominant when issues associated with their specialism become more important. Wallace (1987) found that different communication tactics were used to control specialist contributions. His observations of construction design team interaction using a coding system adapted from Bales' IPA found that group participation is a function of the group characteristics; with participation varying in relation to the way the group develops.

During group interaction, actors' participation is regulated by the feedback or response received to their contribution. Group norms and regulatory mechanisms help to suppress and encourage participation. The amount an individual contributes to discussions and participates within the group is controlled by the behaviour of other group members. Dominant individuals will provide many of the back-channel signals, but other members, being aware of the group norms, may also provide signals that suggest whether continued interaction is acceptable.

Types of feedback or response signals include turn-taking signals, attempt-suppressing signals and back-channel signals. The speaker gives turn-taking and suppressing signals, which are used to defend the right to continue speaking on the same subject or with the same level of emphasis. Back-channel signals are communication acts by other group members, such as agreeing or disagreeing with the speaker. Members may support, disagree with, or even attempt to interrupt the speaker through back-channels, which may help to elaborate the discussions or conversely distract the speaker from the main point. The types of signals and the rate at which they are used relate to the underlying group process, particularly the group regulatory forces. The person sending the prominent back-channel signal will (or will not) also receive support from other group members through sub-level back-channel signals. As well as verbal utterances, facial expressions and body language can convey information and emotion that help control, regulate and direct an individual's interaction. Back-channel signals can be discrete, subtle or overt. Figure 4.1 provides a schematic of the turn-taking dynamics. Influential individuals who have developed communication traits that are capable of changing participation rights can change group regulatory forces; however, the group response is an important determinant of who controls the meeting.

Drawing on group interaction research and the relationship between long-term and short-term agreement, Meyers and Brashers (1999) stated that groups use a form of participation reward system. Those who are co-operating with the group receive communication-helping behaviours and those in competition receive communication-blocking behaviour.

Participation is dependent on the groups' regulatory framework, its influential members and the individual speaker's ability to gain the floor. Participation is also related to a participant's willingness to speak in the group, a factor that may be outside the direct influence of the group process and development (Wallace 1987). An individual's reluctance to communicate in a group may affect the group's participation process. Burke (1974) posits that a person's willingness to communicate accounts for most of an individual's participation during group interaction, assuming that communication takes place in a democratic group environment. People who avoid communication are termed 'reluctant communicators' (Wadleigh 1997). By using a combination of self-report measures and trained and untrained observers, McCroskey (1977, 1997) found that the participant might feel shy due to communication discomfort, fear, inhibition and awkwardness. Through a review of group research that had used the IPA method McCroskey (1997) proposed a 'willingness to communicate trait', where some people will initiate communication and others will not under virtually identical situational constraints.

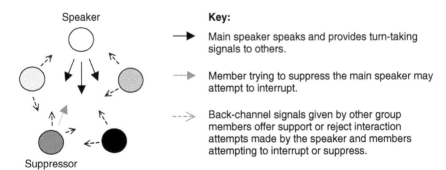

*Figure 4.1* Turn-taking dynamics: Suppressing, supporting and back-channel signals

In a group environment apprehensive individuals talk less than other members and tend to avoid conflict. They are perceived more negatively and are less liked by members who are not apprehensive about communicating (McCroskey and Richmond 1990; Haslett and Ruebush 1999). Highly apprehensive people also have a tendency to attend fewer meetings, avoiding situations where group inter-action is encouraged or seen as the norm (Anderson and Martin 1995; Anderson *et al.* 1999). However, Anderson *et al.* (1999), reporting on the findings of group research, found that the degree of communication apprehension diminishes with group experience.

Members who are reluctant to interact tend to make fewer comments, do not communicate their opinions, ideas or suggestions, and tend to avoid disagree-ment. Although they may have little influence on the direction of the group, their presence may help support the group direction. When group members do not openly disagree with the speaker others may be fooled into thinking that those withholding comments are in agreement, even when they are not. If an apprehensive actor feels that the majority of the group favour the suggestion made, they may choose to withhold their view and accept the direction, while privately disagreeing. Conversely, those members who interact more frequently increase their ability to make others aware of their opinions, ideas and beliefs and have a greater opportunity to influence the direction of the group. Confident speakers are more likely to disagree, challenge others, and defend their own argument; subsequently they have greater opportunity to influence the decision-making process than those who are anxious and less vocal. In decision-making groups, those who talk the most 'win' the most decisions and become leaders (Bales 1953), unless their participation is excessive and antagonises the other members (Hare 1976). Reichers (1987) proposes that more proactive interactors have a greater influence on socialisation and the development of group norms.

Most groups make one or two attempts at encouraging reluctant communic-ators, but then the confident actors choose to carry on discussions paying little attention to less active members. Gorse and Whitehead (2002) found that some group members are good at encouraging the less active to participate. Rather than making one or two attempts to include reluctant communicators the influential and active members made continued attempts using different methods, styles and approaches. Active members used questions, prompts and gestures that provided specific openings for the individual. In a few instances apprehensive members became quite active. Members skilled at encouraging the apprehensive individual to participate were also good at interrupting and controlling the participation rights of the same individual if discussions became prolonged.

## Leadership and participation

Reluctant communicators are unlikely to hold influential positions or be perceived by others as leaders. Relationships have been found between perceived leaders and high levels of verbal participation. Mullen *et al.* (1989), Bales and Slater (1955) and Kirscht *et al.* (1959) found that the person perceived to be leader by group members and observers is the most frequent contributor, being responsible for 50–70 per cent of the participation in groups. One of the problems with the theories that link the group process and leadership is that the studies draw on the perceptions of the group members and their perceptions are often coloured by dominant individuals. Considerable research has shown that those who dominate interaction draw attention to themselves. When asked to nominate group leaders, members often attribute leadership to the most active. Dominant members may be attributed with responsibility for the actions of the group yet have limited effect on resolution of tasks or problems. Bales (1970) found that talkative group members draw attention to themselves through their domination of interaction. This attention may result in other members attributing leadership to that person. Although some actors may dominate conversations, their continuous talkative nature may frustrate, irritate or annoy others and the dominant member's ability to influence group decision would be reduced.

Talk time seems to have a strong link with perceived leadership, although there are mixed views over whether there is a relationship between leaders and their use of interaction content (Pavitt 1999). Using the IPA method Lonetto and Williams (1974) detected no significant differences between the perceived leader and the other group members' proportional use of the interaction categories. These and similar research findings have been used to imply that leaders may be saying nothing different to the other group members, just more of the same. Other researchers have reported differences between high participators who were, and were not, viewed as leaders (Pavitt 1999). Goleman (1996) posits that individuals who fail to use appropriate emotion to emphasise communication will not gain the attention of others, may have difficulty forming relationships, and will have difficulty influencing other actors. Trying to bombard people with constant conversation may gain early attention, but it is unlikely that this will always result in the dominant communicator being nominated as leader. Influential members often realise that those making the most noise have little relevant to contribute and make efforts to encourage the reluctant communicators to participate (Gorse and Whitehead 2002). Before we reject the dominant theory, the turn-taking norms of the group must also be considered. Even if a person wishes to dominate discussions, suppression and back-channel signals may be so strong that only those with considerable communication skill and influence emerge as the dominant communicators. The attributes of dominant communicators may be closely associated with those of leaders.

Bales' (1953) early studies of those perceived as leaders by the group found that differences existed between task and relational leaders, based on their talk time and content. Kirscht *et al.*'s (1959) examination of leader communication, using content analysis and group member perceptions to identify leaders, found that leaders offered more answers than other members. Such findings do not support the view that leaders are simply saying more of the same thing. A problem with these studies is that they are based on internal group perceptions and they do not distinguish between leaders who enable groups to work more effectively than others.

Armstrong and Priola (2001) investigated how people who had a tendency towards analytical and intuitive approaches (cognitive behaviour styles) used the Bales' task and emotional categories under simulated 'work' conditions. Their research involved 100 students working in 11 teams, who were required to produce written responses to a tender application. The students worked in groups, discussing surrounding issues and producing the tender. Using Allinson and Hayes' (1996) cognitive-style index (a self-report questionnaire) team members were assessed on their intuitive and analytic dimensions. They found that there was a tendency for teams to select intuitive individuals as team leaders. The intuitive subjects were more likely to use emotional interaction and initiate socio-emotional behaviour. Intuitive members were more likely to engage in acts of solidarity, being supportive of others, showing affection and attraction. Such behaviour helped maintain relationships with other members. It was proposed that their tendency to use supportive behaviour and engage in socio-emotional exchanges was the primary reason why they tend to be more liked and respected. However, it was also found that the intuitive individuals also engaged in greater task-oriented behaviours than the analytical individuals, which was contrary to Armstrong and Priola's predictions. Earlier research suggested that intuitive individuals would use less task-orientated behaviour than analytic members, although such behaviour is suggested to be context dependent (Gruenfeld and Lin 1984). The earlier research on which the Armstrong and Priola hypothesis was based had been conducted in well-structured environments. Armstrong and Priola considered their context to be less structured and more natural. Further investigation is required to determine whether the findings of the laboratory work correspond with the behaviours of construction participants in real life.

## Task and relational interaction

Both unidimensional and multi-dimensional schools of thought pay considerable attention to the relationship between task and emotional exchanges during group communication. Scholars generally agree that the two main interrelated dimensions of group life are task and social dimensions (Frey 1999). Work-oriented groups need to maintain a balance between task and social demands (Frey 1999; Keyton 1999, 2000). Bales (1953, 1970) found that as groups address problems, emotions start to develop and, as a result of disagreement, tension is built up between members as they focus on the problem rather than relationships. Bales' observations noted that conflict, even when constructive, leads to tension that can damage the cohesiveness of the group and threaten group maintenance; yet too much attention to cohesion tends to stifle constructive conflict and threatens the group's ability to solve problems. In order for a group to be effective task issues must be discussed, conflict will emerge, and relationships must be managed so that they are sufficiently sustained to bring the discussion to a successful conclusion. Cline (1994) identified the importance of functional conflict to avoid 'groupthink' and improve the decision-making process, but conflict may also damage relationships between group members if not managed with sensitivity.

Conflict often emerges from perceived failure. If group members fail to meet their level of aspiration they may or may not try harder. Moderate levels of failure have been found to produce greater effort towards organisation than either low or high failure levels (Hare 1976). When the failure level is high there is a greater chance that people perceive the task to be impossible or the chance of failure to be so high that it is not worth the effort. Very low levels of failure may be taken as a job achieved, but not done too well. Moderate failure, where the task is considered tough but achievable, may present a challenge and a chance to prove to oneself and others that the task can be achieved. Thus, moderate levels of conflict (being related to perceived failure) may be productive. Negative feedback can be stressful (Shapiro and Leiderman 1967) and group members need to be aware of the development of socio-emotional tension. Debate and negative emotional exchanges may threaten relationships; however negative emotion can be useful if positive socio-emotional interaction is also used to maintain relationships.

Any socio-emotional tension that develops is removed by positive emotional acts (such as showing support, joking and praising) and negative emotional acts (such as disagreements, an expression of frustration and even aggression). Bales found that if socio-emotional issues are not addressed in a timely fashion the increase in tension may inhibit the group's ability to progress its work. Bales states that groups must maintain their equilibrium, moving backwards and forwards between task and socio-emotional–related issues. Negative and positive socio-emotional interaction is interlinked with the group's task-based interaction. The task-based interaction will provide information about possible ideas, action and direction of the group. The positive and negative emotional signals will provides clues about how group members feel about the suggestions and encourage others to provide further information. Through suggestions and ideas, the subsequent testing of them and the reaction of the group (conveyed in rational and emotional responses), members are able to learn what is acceptable behaviour for the group.

Too much attention to task interaction limits the communication required to build and maintain relationships. If groups are to perform effectively positive reinforcement (agreeing, showing solidarity, being friendly and helping release tension) is needed to offset negative reactions (showing tension, being antagonistic, appearing to be unfriendly and disagreeing). Negative emotions should be expressed, although the impact of the negative act needs to be reduced or tempered with statements that reinforce the importance of the relationship. Subtle positive socio-emotional acts such as agreeing, joking, smiling, laughing, praising, showing support may help reinforce the relationship.

Clearly it is important to focus on the task and disagree if necessary, but this should not be at the expense of positive emotional exchanges that maintain relationships. Drawing on McGrath's (1984) review of small group communication research, Keyton (1999) suggests that positive relational acts need to be in excess of negative relational acts to accomplish tasks successfully. It is suggested that a larger positive to negative ratio facilitates and regulates the flow of interaction among members and contributes to the motivation and satisfaction of the members (Bales 1953; Keyton 1999).

Bales (1953) found that after group members had dealt with a problematic task they would diffuse negative emotions with positive emotional discourse, returning to the task issues once the tension had been dissipated. When negative socio-emotional interaction occurs the tension is released in stages, first through task-related discussion, then via positive socio-emotional exchanges. The total communication of healthy groups is said to contain several times as much positive socio-emotional acts as negative socio-emotional acts and about twice as much task-related communication compared to maintenance acts (Shepherd 1964). Maintenance acts being the combined total of positive and negative socio-emotional acts.

Group members feel comfortable in a positive socio-emotional environment. Members prefer positive feedback (Jacobs *et al.* 1974), and interaction that suggests the group is effective can help to increase morale (Frye 1966; Streufert and Streufert 1969), but too much may be counterproductive. Cline (1994) used Bales' IPA to analyse transcripts of groups that were known to have made flawed decisions and found that too much emphasis on agreement resulted in unsuccessful outcomes. In 'real life' groups that were found to be associated with groupthink, the levels of agreement were ten times higher than disagreement and, when groupthink occurred in laboratory studies, agreement was seven times greater. Although groups that avoided groupthink used more positive than negative socio-emotional interaction the difference in levels of agreement to disagreement was much lower (ratio 5:1) than unsuccessful groups. Thus, while the results of laboratory experiment groups suggest agreements should be several times higher than disagreements if members are to be 'satisfied', such a high ratio could be detrimental to the performance of workgroups and may result in groupthink.

Bales (1950) also identified that some socio-emotional acts help to develop understanding, for example when talking, communicators acknowledge understanding by emotional expression. Bormann (1996) and Trenholm and Jensen (1995) suggest that when group members respond emotionally to a dramatic situation they are openly proclaiming commitment (or not). Such expressions strengthen the group's social system; group members develop an understanding of how others will react to future situations. Such behaviour structures the group's decision-making.

The distinction between task and socio-emotional behaviours still remains a fundamental assumption of group communication research (Poole 1999). However, rather than the relationship between task and socio-emotional behaviours, proposed by Bales in his equilibrium theory, there has been a tendency for scholars to believe that task and social dimensions are in competition (Frey 1999). The result is that many studies have a greater emphasis on the task-based factors, neglecting social relationship issues (Frey 1999; Keyton 1999). Keyton (1999) points out that even when research does consider emotions, they are usually considered with respect to their impact on task messages or outcomes rather than their impact on relationships. In many studies, relational acts are often found to facilitate the group development process but inhibit group performance

(Keyton 1999). However, recent research has identified that relational communication matters on many levels, including its role in establishing the climate and structure within which group tasks are accomplished (Keyton 1999). The role of task and relational communication appears to be more interlinked than previously thought.

It is clear that the use of socio-emotional interaction is important in the accomplishment of tasks. The combined effort of the group requires exchanges that develop and maintain the group and provides a social structure capable of decision-making and task accomplishment. The information gleaned from the use of Bales' IPA can be used to estimate the effectiveness of a group as well as individual member's performance, allowing members to change those communication patterns that hinder performance (Schultz 1999). Use of socio-emotional communication, which helps us to understand the meaning and value of the task message and which can also threaten and reinforce relationships, is fundamental to group research.

## Task and relationship roles within groups

Referring to work by Bales (1950, 1953, 1970) and theorising about group performance, Pavitt (1999) suggested that members must undertake roles to ensure that task and maintenance goals are maintained. Concentrating on these roles Bales (1950) found that those judged to hold positions of leadership had certain tendencies. The most frequent talker tended to be the most highly respected but most disliked member of the group; this role is often referred to as the 'task leader'. The next most frequent talker was not as respected as the first but tended to be the most liked member of the group; this member has been labelled as the 'maintenance leader', the person who was responsible for supporting the task leader and other members of the group. The maintenance leader ensures that the group relationships are sustained during difficult periods. By lubricating the task discussion with supportive interludes, positive socio-emotional relationships are maintained. Focus is on controlling the interaction so that tasks can be discussed and relationships sustained. The maintenance leader will be involved with the task-based discussions, but is aware of the group's emotional needs.

A study of interaction leadership traits by Heinicke and Bales (1953) found that individuals who held high status would participate and contribute the most during early meetings, with their contribution reducing in subsequent meetings. Laboratory studies undertaken at Northwestern and Harvard University comprised six groups of five students each, which met once per week for six consecutive sessions. At each session, groups were given a different problem and asked to write an advice letter or proposal stating how to overcome the problem. Each session was audio recorded and observed through a one-way mirror. Observers ranked the member's displayed leadership. In controlled experiments undertaken prior to the main research it was found that ranking of leadership by group members and observers had a strong correlation. The status ratings were based on 'who had the best ideas', 'who did the most to guide the group' and 'who stood out the most definitely as leader'. Regardless of the differences in the location of the groups and the number of sessions attended, the trends found were the same for the two sets of groups. Those subjects who were ranked high on the status rating initiated greater amounts of opinions and suggestions to the group, offering direction, and also received a greater amount of agreement from other group members. An examination of high-consensus and low-consensus groups found that leaders in high-consensus groups used even greater amounts of directive communication. Conflict between high and low status members was found in both groups, although it reduced in high-consensus groups. The successful groups used greater amounts of information and suggestions to guide the group while also engaging in socio-emotional exchanges. Less successful groups offered less direction and were unable to overcome disagreements and conflict that emerged from information exchanges.

At first, Bales (1950) failed to examine the extent that leaders were associated with task and maintenance functions. The second problem that emerged was that the participants' ratings implied a distinction between two different types of task leaders; the best-ideas person (also known as the substantive leader) and the person giving the most guidance (the procedural leader) (Pavitt 1999). The overall group leader was found to be procedural rather than the best ideas person (Bales and Slater 1955). The use of suggestions and directions emerged as an important aspect of successful groups in all of these studies.

While the split between different types of leaders in a group has been questioned, the functions of leaders have largely been substantiated. Pavitt (1999) noted that the distinction between task and maintenance leadership functions as well as the further divisions of the task function into substantive and procedural appears to be sound, although it has been criticised. Wyatt (1993) found that in certain situations, such as therapy and support groups, the task is to build relationships, task work being the same as maintenance, and a communication act could actually serve both roles. The extreme role differentiation between task and maintenance leaders sometimes appears artificial (Pavitt 1999), as such roles often change between meetings (Turk 1961). In Wallace's (1987) study of the construction design team, the group used social and emotional interactions to nominate and elect the group leader and support their status. As the group tasks changed, socio-emotional interaction was used to remove those elected, enabling others to become more influential.

Task-based communication is often used to steer the group towards the group goal, but it is really just a label for communication that serves a function and does not contain any emotional content. When communication acts use emotion they should be classed as having either positive or negative socio-emotional context. In certain contexts, such as therapy or support groups, it would be expected that there is a tendency towards greater use of socio-emotional communication, and while this would help achieve the group goal it would not be classed as task-based communication. Communication acts that fall into socio-emotional categories should be classed as such, thus Wyatt's conclusions are slightly misleading. The task of undertaking therapy is emotionally intensive and achieved through high levels of socio-emotional interaction. One may be tempted to identify such acts as contributing to the task in hand, but they are socio-emotional acts and signals used to establish and maintain a relationship between counsellor or therapist and client. Studies that recognise that there is a tendency to use more, or less, emotional communication in certain situations should also examine what effect the communication pattern has on the participants and what types of behaviour tend to improve the group experience or product.

The Bales system is not used to capture the detail of the task, but is used to identify the communication act and its overarching nature. Many communication acts have two or more purposes. The Bales system captures that which is most obvious to the observer. The observer uses their emotional sensitivity to capture signals of a socio-emotional nature. Certain environments may be particularly socio-emotionally intensive or, for that matter, heavily dependent on task-based interaction. To further understand the nature of emotionally intensive environments further research methods may be required in addition to the use of IPA.

## *Observations of socio-emotional and task-based interaction in specific contexts*

Early studies of group behaviour (Morris 1966) using IPA found that groups exhibited regular patterns of interaction that were specific to the context in which they were observed. When the context of the group setting is very controlled, small changes in the group size do not have a profound effect on the behaviour of the group; for example Bales' (1950) observations recorded little difference in the interaction patterns of different-sized groups of chess players (Figure 4.2). However, quite a large proportion of problem solving behaviour is relatively consistent regardless of setting (Mann 1961). For example, Bales' (1970) study of adult norms generated from 21 studies of different groups (Figure 4.3).

The characteristics of the groups (Manheim 1963) and the reason for discussion or nature of the problems discussed can affect an interaction (Bales 1950). Differences between groups may be subtle, yet influential. Small differences in socio-emotional behaviour have been found to produce significant changes in the groups' behaviour (Hare 1976). Bales investigated a number of groups with differing characteristics, problems and tasks to investigate these variables. The interactions of adult groups were found to be different from groups of children. As the children developed, their interaction patterns changed. Preschool interaction was found to be unstructured, and high levels of emotional interaction were observed, the interaction of the child being described as unhindered, uninhibited and uncontrolled (Figure 4.4). The children's emotional communication would quickly change depending on their immediate needs. Children quickly show their frustration and express satisfaction; adult behaviour is more inhibited and controlled. The difference between children and adults is important as is the way children and adults use task-based information. As adults we have learned

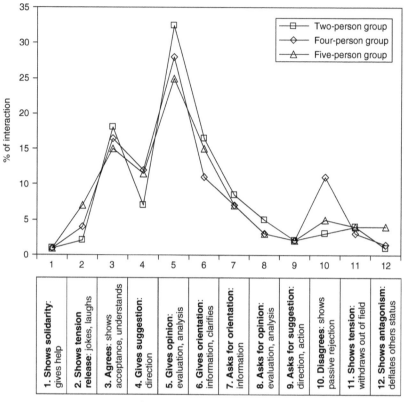

*Figure 4.2* Interaction profiles of different-sized chess groups (adapted from Bales 1950: 149)

to control our emotions, yet subtle use of the same emotions is used to gain influence in relationships and groups.

The suggestions made by the children were almost entirely unsupported by any preliminary or accompanying analysis, inference or persuasion (Figure 4.4). Later studies of older children found that they had developed some control of extreme emotional expression, although the amount of tension shown and tension released, such as by laughing and joking, were found to be much higher than adult groups (Bales 1950; Socha and Socha 1994). It is to be expected that the behaviour of adults is more reserved; however, it is important to consider that this is a developed, learned and controlled response. Adults may not express their emotions as readily as children, but they may still experience the feeling of enjoyment, frustration, being hurt, not belonging etc. Socio-emotional signals may be suppressed when compared with children, but it would be wrong to ignore their function and use within a group. Equally it is important to recognise and

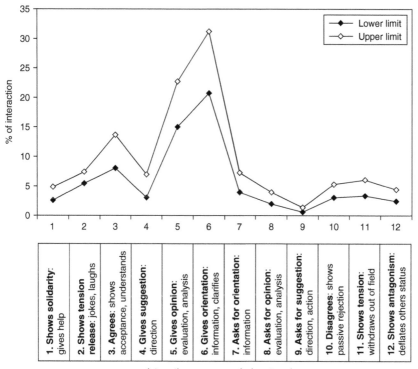

*Figure 4.3* Norms of adult interaction (adapted from Bales 1970: 473): Based on 21
studies

understand how adults develop their task-based interaction in order to convey
information and influence others.

Adult groups do not tend to have high levels of laughing and joking, and tend
to use even lower levels of negative emotional expression (Figure 4.3). The level
of socio-emotional interaction is low when compared to task-based interaction.
The interaction of adult groups is characterised by high levels of information
giving, supported by even higher levels of explanations and opinions, while
asking for information, opinions and suggestions occurs less than half the time in
task-based interaction. The level of information and opinions is higher than that of
suggestions. Before suggestions can be made the context can be set by exchanges
of information. Exchanges of information provide background data that helps
others orientate themselves within the group and the context of the discussion.
Opinions are slightly more sensitive because they associate the person speaking
with the message conveyed, whereas suggestions are much more control orientated. When offering direction an individual gives the direction that they or other
actors ought to take. Suggestion and directions are relatively hard statements,

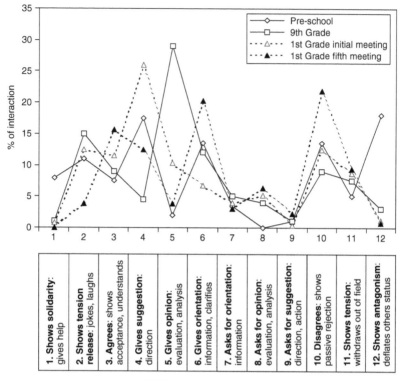

*Figure 4.4* Interaction of preschool and 9th grade children (adapted from Bales 1950: 23) and 1st grade children (adapted from Socha and Socha 1994: 242)

when compared to that of opinions, analysis and information-based acts. Thus, there is a continuum of sensitivity and degree of intrusion that each message potentially conveys on other group members. Information is least sensitive to others, whereas opinions and suggestions become potentially more intensive as they convey self-belief and are intrusive as they have potential to engage others into action, and ultimately direct the group. There is also a hierarchy of socio-emotion communication acts. As a person moves from disagreeing to showing frustration and then to being aggressive the intensity of the message increases.

Bales (1950) made several interesting observations of adult groups. His obser-vations of a group of five married couples made two points. One was that giving information exceeded giving opinion and explanation. Although he found explan-ations to be higher than opinions (Bales 1950) in early studies, later work showed that many adult groups used higher levels of information giving than giving opinion and explanation (Bales 1970). Explanation is normally used most in adult groups to support the understanding of information. However, in married couples it was claimed that the strength of the relationship might be such that information can be presented and any inferences to the nature and meaning of information will, in many cases, be mutually understood. Thus, when relationships are established the degree of explanation needed diminishes rapidly. Equally, when professionals share similar education and industrial experiences their understanding of issues may be sufficiently similar that the degree of explanation needed may be reduced. Gameson's (1992) observations of construction professionals and clients found that the primary category for construction professionals when meeting clients for the first time was giving information, although this did vary between meetings. The second point of interest from Bales' (1950) study of married couples was that the level of antagonism was higher than observed in other adult groups (Figure 4.5). Antagonism in marital relationships may be more marked because of the longevity of the relationship, permitting the use of extreme negative emotional behaviour without endangering the relationship. Thus, the strength of the relationship (or the need to maintain the relationship) may become so strong that members feel that they are able to express greater levels of negative socio-emotional communication in order to shape group behaviour.

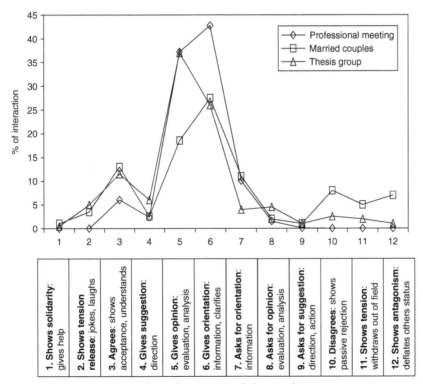

*Figure 4.5* Interaction of married couples and thesis group (Bales 1950: 25) and construction client – professional first meeting (Gameson 1992: 312–317)

Other temporary adult relationships observed by Bales (1950) hardly used the extreme emotional behaviours, such as showing solidarity or antagonism. The level of negative socio-emotional interaction recorded in the study of an academic discussion group was described as a bare minimum, accounting for only 4 per cent of the total interaction observed. Gameson (1992) found virtually no occurrence of the extreme emotional categories in his study of meetings (Figure 4.5), which to some extent supports Heinicke and Bales' (1953) work. They recorded interaction development in two groups of studies over four and six sequential meetings. They found that task-based communication was high in the first meeting and as the groups progressed through the meetings there was a significant reduction in task-orientated categories (giving orientation, opinions and suggestions) and a concomitant rise in the socio-emotional categories (agreeing, disagreeing, tension and support). Negative reactions were low in the first session, rose sharply in the following session and thereafter decreased. Positive reactions increased throughout the meetings, with a substantial rise in the final meeting. There was also a shift from the more neutral task-oriented agreement to the more effectively charged positive reactions. However, when participants are aware that they are to take part in a final group meeting, increases in emotional expression could be due to tension release as members attempt to make their final contribution. Alternatively, it could be a result of camaraderie, as research participants are aware they no longer have to undertake the group exercises, although this is not stated in Heinicke and Bales' findings. Even in laboratory settings it is difficult to know exactly why such behaviour occurs.

Heinicke and Bales' (1953) study also examined the different types of interaction found in groups that had high and low consensus. A participant self-report scale and observer analysis determined the level of consensus reached (a strong positive correlation between the self-report scale and observer analysis was found). High consensus groups showed a significant increase in solidarity and tension release, a significant decrease in the amount of negative social interaction and significant decrease in the amount of suggestions and opinions. It is difficult to know to what extent such behaviours transfer to other contexts. Nevertheless, Bales (1950) suggests that certain interaction patterns may be similar although differences in the group exist; for example, a group dealing with a typical problem in a different context may exhibit similar interaction patterns to other groups addressing the same type of problem.

### Issues to consider in a male-oriented context

Construction is predominantly male orientated and although gender balance is not our main concern, we need to recognise that the type of interaction observed in construction teams may be related to male dominance. Aries (1976) observed the interaction of two male, two female and two mixed student groups who met in five one-and-a-half hour sessions with the task of getting to know each other. Using the IPA, Aries found that males were the dominant interactors in the mixed groups. In the male groups those dominant in the first meetings maintained their high level of interaction throughout all meetings. In the female group there was greater distribution of interaction tendencies. Following the initial meetings those previously dominant encouraged interaction from less active parties and the interaction became more distributed. When compared to the mixed group, members of the male groups addressed more of their communication to the group as a whole. Male members were found to be more competitive and aggressive when with other males, attempting to dominate interaction using jokes and information to establish their status. Females were more inclined to show warmth and friendship in the form of joking and laughter. The differences in interaction style between males and females were less dramatic in mixed groups.

In a later study by Hawkins and Power (1999), 39 male and 59 female students were formed into 18 groups and asked to produce a semester paper, which formed part of their assessment. Contrary to Aries' (1976) results, there were no significant differences in the participation of males and females. Using an adapted version of Bales' IPA, Hawkins and Power found that the use of question asking was different for men and women. Although there was no difference in the level of question asking, females made greater use of probing questions. Such questions are more likely to foster co-operation and connection with others, helping to strengthen and deepen relationships. The nature of such questions was to invite elaboration of arguments, sharing of information and opinions, and inviting increased participation from fellow members. The different interaction styles are consistent with the behaviour observed by Aries (1976). Male-dominated groups are more competitive and less inclined to pursue relationships.

## IPA and multidisciplinary groups

A few researchers that have used the IPA tool refer to 'work groups' in their title or abstract (e.g. Armstrong and Priola 2001), although research using IPA in commercial and industrial environments is scarce. A publication that did report the findings of a multidisciplinary group was Bell's (2001) investigation of 15 child protection teams. Although not a commercial group the teams were made up of different professionals and agencies operating in 'real' meetings. Group members were predominantly female. High levels of task interaction ranging from 83 to 93 per cent typified the discussions observed. When the proportion of giving information, opinions and suggestions were compared with asking for information, opinions and suggestions, all of the professionals except the psychologist gave, rather than asked for, information. Consistent with earlier studies some members were more influential and dominant than others and interaction within the groups was not evenly distributed. Where an agency involved in the meeting was represented by more than one member the senior representatives would make a greater contribution than the less senior representatives. Interaction was not evenly distributed across the group and would be dominated by one or two members. As the group sizes increased the proportion of members contributing to the meeting decreased, this is consistent with the reports by Bales (1950, 1970). In three of the teams observed, 21 per cent of the members communicated less than 1 per cent of the group interaction. Bell concludes that the lack of contributions made by many of the specialists meant that the multidisciplinary teams failed to provide a holistic view necessary to determine the needs of a maltreated child. Even though this context is somewhat removed from the multidisciplinary nature of construction teams it provides a perspective on the workings of a multidisciplinary context, and it also raises a question as to whether specialist perspectives are fully utilised in multidisciplinary team meetings.

### Effects of closure on task and social interaction

Problem solving during the construction process is subject to the pressure of closure. Tasks must be completed by a set time and on completion the project comes to an end. Meetings also have timescales and must be brought to a satisfactory close on time. Closure is defined as the desire for a definite answer to a question, issue or problem (Kruglanski 1990). Problems experienced in construction projects must be resolved if the building is to be completed within the scheduled duration. The need for closure is increased when time constraints are imposed (De Grada *et al.* 1999). In research involving 24 four-person student groups, engaging in role-play discussions to resolve fictional corporate problems, significantly different interaction patterns were found when the groups were subject to time constraints. Using Bales' IPA, De Grada *et al.* found lower levels of positive socio-emotional interaction and a higher occurrence of 'giving' task-type interaction in groups that were subject to making decisions within a time constraint. De Grada *et al.* concluded that the time pressure prevented the groups engaging in 'social niceties', resulting in groups emitting a lower proportion of positive socio-emotional acts. Although De Grada *et al.* attempted to simulate a commercial environment by asking students to engage in a business-related problem, a weakness of the findings when considered against a real business environment is that the students only took part in one meeting. The time-constrained meeting lasted 50 minutes and the unconstrained meeting lasted 66 minutes. In real commercial projects, while constrained by time factors, the professionals would normally make decisions over a series of meetings. They also have to sustain relationships sufficiently so that they can work together during future meetings. The lower level of socio-emotional interaction observed in the group subject to time constraint is still interesting. Bales (1950, 1970) claims that positive socio-emotion is used to disperse tension that results from task-based discussions and to maintain relationships. If the participants had been observed over a series of time-constrained meetings interaction patterns may have been different. Although socio-emotional interaction was less in the time-constrained situation, both sets of results produced higher levels of socio-emotional interaction than many of the earlier studies.

De Grada *et al.*'s profile of group discussion differed from the classic data described by Bales (1970), with levels of positive and negative socio-emotional behaviour being considerably higher than the norms of adult interaction proposed by Bales. De Grada *et al.* suggest that the high level of conflict and positive social interaction could have been due to the bargaining nature of the simulated discussions, which might have been more emotionally involving for participants than the problem solving tasks used by Bales. The time constraint also encouraged a conversational pattern wherein some members manifest greater dominance of the discourse than do others. Using scales to assess individuals' tendencies towards closure, De Grada *et al.* found that those with a greater tendency towards resolving problems quickly were more dominant in time-constrained discussions. They also tended to adopt more autocratic styles of leadership, giving directions to other members of the group.

## Social influence and persuasion

A few works have examined the way in which individuals or groups use communication to influence others and gain compliance. A factor affecting social interaction is the power of each side to affect the other (Patchen 1993). Frost (1987) used the term 'surface power and politics' as a label for social influence that is used by a person to get what they want from a decision, negotiation or discussion. It is suggested that people who have, or gain, greater power, use coercion, whereas those with less power tend to submit when a more powerful adversary uses power against them (Patchen 1993). Hare's (1976) review of communication studies suggests that regardless of whether a person's power is based on legitimacy, ability to co-ordinate group activity, skill, or some other factor, the more the person attempts to influence another person the more likely he or she will be successful, especially if the recipient is willing to accept the proposition and peers do not set counter norms.

Using discourse analysis to examine a city council meeting's transcripts, Barge and Keyton (1994) found that politicians used argument as a means of persuasion. During an observation of a council meeting, an autocratic mayor's power was reduced by analytical arguments that helped to build relationships and alliances with other members at the meeting. To create and sustain a context for his proposition the Mayor of the council adopted an autocratic position claiming outright authority for his argument based on position. He made an assertive claim that his elected position gave him outright authority to decide what was discussed at a council meeting. A council member challenged the proposition as undemocratic. The Mayor reinforced his position, becoming even more assertive. The council member repeatedly challenged this position by establishing a rapport with those who were aggrieved by the Mayor's stance. Thus, not only was the Mayor's attempt to dictate action challenged by the initial disagreement, but the challenge was reinforced by support from others.

The council members and the Mayor used various influence strategies to gain control over meeting discussions. The Mayor used simple arguments (hard control statements proclaiming his supposed authority and power) with minimal explanation as a method of claiming authority to control discussions. However, when unsubstantiated assertions were made (by the Mayor), another member attacked the way the statement and position was set out. By open analysis of the statements made by the Mayor the member in opposition was able to use the autocratic position adopted by the Mayor against the Mayor. Attacking the unilateral nature of the statements made and forming relationships with others achieved the challenge. The council member who was observed to have the most influence challenged and insulted the Mayor, questioning whether anyone who could assume such authority could operate in a democratic way. Others were able to use their elected positions to exercise their own opinion. Through unsuccessful use of the simple 'hard-line' approach by the Mayor, and successful use of insults, and by gaining support from other members, rather than insisting on support, a coalition formed against the authoritarian 'hard-line' attitude. Although De Grada *et al.* found autocratic behaviour to be more influential in time-constrained situations; in this case study Barge and Keyton show that the extremes of dominant autocratic behaviour prove unsuccessful if relationships are not maintained. Using suggestions, giving direction, asking questions and disagreeing are repeatedly observed during interaction of members who are perceived to influence group behaviour. Control statements and negative and positive socio-emotional exchanges are clearly an important mechanism for ensuring information is considered properly by groups.

The common thread through the literature on group work shows a link between the way task-based information is used to provide information, analyse situations and direct the group and the need for socio-emotional interaction as a decision-making and support tool. Socio-emotional interaction provides clues and signals that indicate whether task information is being accepted, equally emotional responses can be used to support and contest others.

### Influence and persuasion in the business context

Influencing skills enable the development of relationships (Craig and Jassim, 1995) and are a key skill of good managers. Godefroy and Robert (1998) claim that people with a high persuasive ability can use their skill to handle conflicts constructively and hence to promote openness and constructive debate. To perform effectively construction project managers need the ability to negotiate and persuade others to take action (Hodgetts 1968; Sotiriou and Wittmer 2001). Creyer (1997) suggests that prior to initial negotiations there is a small level of social persuasion which may influence the way negotiations are actually started, potentially providing one of the parties with the 'upper hand' before real negotiations have commenced. To develop a social link the negotiator needs to tentatively get to know the other party, making comments which would not offend and drawing out background information from the other party so that a relationship can be developed without any initial upset. Such encounters rely on information giving and seeking and on low level positive socio-emotional reinforcement.

Morley (1984) suggests that one way to view negotiation is as a struggle. This approach emphasises concealment and competitive tactics. An alternative view of negotiation is as collaboration, a process in which parties make sacrifices rather than demand concessions in the pursuit of some overriding goal (Strauss 1978). Strauss divides negotiations into those that are competitive and those that are collaborative, whereas Scott (1988) presents two contrasting styles, which are labelled 'competitive' and 'constructive'. Pruitt (1981) discusses competitive and co-ordinative negotiating tactics and the concept of integrative bargaining. He defines 'integrative bargaining' as the search for mutually beneficial agreements. Pruitt (1981) adds to this by describing how this might involve working towards novel or unique outcomes that could produce considerably greater benefits for both parties than a straight compromise. It is suggested that, where parties use collaborative strategies during negotiations, it is more likely that a mutually agreeable solution will be produced (Harwood 2002). Both the collaborative and competitive styles of conflict involve positive and negative socio-emotional discussion. The collaborative method would have a tendency to maintain and repair relationships during discussion. The competitive style would be prepared to threaten the relationship and have a greater tendency towards extremes of the negative socio-emotional traits. Both approaches would also use task-based logic to explain the rationale; however, getting mutual understanding would be far more important in the collaborative approach than the competitive approach.

In situations where negotiators or project managers feel that they have to satisfy tough demands (e.g. negotiations with stakeholders within a project), they will adopt a more competitive approach (Harwood 2002). However, Murray (1986) noted that the use of a competitive approach emphasises emotional pressure. Emotional pressure, especially if tension is not managed, may lead to distorted communication and result in misleading information. Subsequently, the chances of a project manager achieving success may be hindered.

Yukl (1998), Gupta and Case (1999) and Harwood (2002) all found that different influencing tactics were being adopted at different stages of negotiations. The level of resistance encountered by the negotiator often governed the type of influencing technique. Yukl (1998) suggested that initial influence attempts often involved a simple request through a relatively weak form of rational persuasion, using task-based logic to explain the situation. However, Yukl goes on to say that if the project manager or negotiator feels that his or her attempts are failing, or just falling short of that required, then other types of soft persuasion tactics will be introduced, such as, personal influence, appeals, ingratiation, consultation or inspirational tactics.

Redmond and Gorse (2004) undertook a small study to observe and identify the external influencing and persuasion techniques used by business bank managers during negotiations. Managers were observed having face-to-face meetings with potential clients, during which negotiations took place regarding possible financing of a new contract. The study was based on four senior business managers from a major bank and their attempts to persuade various outside contacts, including financial directors and property developers to engage in business relations. All the business managers were involved in initial negotiations of new projects that required funding.

The research attempted to establish what influencing tactics, methods or techniques formed part of a successful or unsuccessful meeting. When subsequent meetings or negotiations took place between the bank and the potential client, the meeting was considered a success. Audio recordings of the meetings were taken between the finance directors or property developers and the business bank managers. The recordings were then categorised using the Bales' IPA technique and discourse analysis. Based purely on the Bales' IPA information, the two successful negotiations had similar start patterns. On both occasions the banker consciously or unconsciously started the meeting by raising the status of their opposite members, while at the same time offering passive acceptance and agreement of the information being presented. Both of these tactics were adopted from the beginning and continued for at least the first quarter of the meeting. These tactics were accompanied by the offering of information or by requesting information, which was forthcoming.

The successful negotiations showed traits that were comparable to the collaborative approach of negotiation. Here, the two parties readily exchange information and seem willing and comfortable with the sharing of information. This is similar to successful negotiating tactic observed by Hayes (1991), Strauss (1978) and Harwood (2002). Both negotiators portray attributes of joint problem solving (sharing information and ideas). Hayes (1991) believes that solving problems together will lead to a more beneficial solution for all concerned. The collaborative method was accompanied by the use of ingratiation-influencing tactics in the early stages of the meeting, which in some instances was reciprocated by the opponent suggesting ideas and changing those ideas to accommodate others beliefs.

The exchange tactic was also used in the two successful negotiations (it is noted that the Bales' IPA system was not used in this part of the analysis). In this instance the rewards for either party came about via the negotiations and the sharing of information that was of value to the other party, at which point benefits and rewards were identified and presented by each negotiator to their respective opponents.

From this point in the successful negotiations, the bankers employed the use of personal appeal tactics. The bankers, having first identified benefits to the opponent, which may not have been recognised, adopted the use of the personal appeals. This tactic is usually employed by opponents that are familiar with or friends of each other, or by one person on another to test the strength of trust or level of friendship developing (Hayes 1991; Siminitras and Thomas 1998; Yukl 1998). Such interaction makes use of socio-emotional categories to reinforce the relationship. Based on the previous research, an influencing tactic is more likely to be successful if the opponent perceives the tactic to be a socially acceptable form of behaviour. This would mean that no one tactic or technique used in isolation would prove successful or unsuccessful. Research suggests that external influencing and negotiation is a combination of techniques (Siminitras and Thomas 1998; Yukl 1998; Gupta and Case 1999).

Rational persuasion is often used during negotiations. The most common form of rational persuasion consists of logical arguments and factual evidence to help develop understanding and explain the situation. Harwood (2002) suggests that rational persuasion is most appropriate when the opponent shares the same task objectives, but does not recognise the proposal in its current format as the best way to attain the objectives. Both Yukl (1998) and Harwood (2002) recognise that in a situation where the opponents have incompatible objectives or end goals, this type of influencing tactic is unlikely to be successful for obtaining commitment or partial agreement. In the two unsuccessful negotiations, it was also found that the effectiveness of rational persuasion diminished when it was used alone or with harder influencing tactics such as pressure and aggression tactics. The results help to illustrate the use of different communication acts and their use in different influencing and persuasion approaches.

## *Disagreement, argument and aggression*

Disagreement is often seen as a negative term (Cline 1994), yet it is found in most observations of group interaction (Bales 1950, 1970). Moreover, Cline's agreement–disagreement analysis of three transcripts that were used by prosecutors in the Watergate case, using an adapted version of IPA, found that when groups avoid disagreement the vulnerability of a proposal may be overlooked. In the Watergate case, participants that were locked into a groupthink frame of mind showed a higher ratio of agreement to disagreement than non-groupthink discussants. Groupthink occurs where individuals, within the group, do not agree, but fail to show their concerns because of group pressure or they believe the majority of the group are in agreement. Ball's (1994) investigation of conversations and reports of the J.F. Kennedy administration using symbolic interaction analysis found that the President fostered norms of secrecy that prevented and suppressed challenges and conflict. The lack of conflict and challenge subsequently led to confusion during the decision-making processes. Conflict during discussions can have positive effects on decision-making, challenging and evaluating proposals and exposing risks of decision. However, if conflict develops into a full dispute (allocating blame and fault), outcomes of a satisfactory nature are substantially reduced, as the relationship is threatened.

A certain amount of challenge, evaluation and disagreement is necessary to appraise alternatives and reduce the risks. Furthermore, Averill's (1993) review of anger-based research found that a typical angry episode would often result in change which had positive benefits, and typically the relationship within which the anger was expressed was strengthened more often than it was weakened. Emotional expression helps others to recognise an individual's preferred beliefs, behaviours and actions. It exposes the structure and routine in which the individual would like to work. In groups such expressions are needed to establish what is acceptable and what is not and in which areas conflict is likely to occur. The decision-making structure of a group is dependent on individuals expressing their agreement and disagreement and working through any variance. As individuals work through their differences they develop a much deeper understanding of others' beliefs and values. However, people may choose to avoid disagreements to enable them to pursue relationship goals, believing that disagreeing would weaken the relationship (Wallace 1987; Cline 1994). Clearly, it is important to expose differences of opinion and explore issues in depth, but it is equally important to identify and respond to an individual's values and beliefs. During difficult discussions it is important that tension is reduced. Communication acts that offer support and release tension can help to sustain and rebuild relationships.

A distinction has been made between argumentative and verbally aggressive behaviours. Argumentativeness involves making refuting statements, whereas verbal aggressiveness involves attacking the self-concept of another (Anderson *et al.* 1999). Humans are extremely sensitive to negative emotional and aggressive behaviours, and we seldom experience difficulty recognising when we, or others, are being aggressive (Averill 1993). Our evolution and survival is partially based on an ability to recognise aggression and respond to it by escaping from, or confronting, the situation. Our sensitivity to negative emotional expression tends to be very sharp; actors can quickly distinguish between concern, distress, disagreement, frustration and aggression and take appropriate actions. Some actors are more inclined to take part in altercations or to help others than their colleagues who may choose to avoid emotional engagement. Goleman (1996) associates this difference with our emotional development and emotional intelligence.

Research has shown that members who are argumentative express greater satisfaction with communication, and perceive their groups as reaching higher levels of consensus and cohesion, than the members who are not argumentative but are verbally aggressive (Anderson *et al.* 1999). Argumentative members make contributions that are more rational and thorough than their less argumentative counterparts. Verbally aggressive members alienate group members. Mild forms of aggression, such as threatening to break off talks, committing and sticking to one position, imposing time pressures on opponents or belittling the opponent's argument, are often used in negotiations (Pruitt *et al.* 1993).

### *Agreement*

Few studies have investigated the nature of agreement within groups, and even less have considered what happens when group members seem to agree but probably do not. Research that did consider this phenomenon was that undertaken by Cline (1994). The research, based on the tapes and transcripts from the Watergate case using IPA and content analysis found that much of the agreement within the group discussions lacked accompanying substance. The level of agreement found in the Watergate case (29 per cent of group interaction) was higher than previous investigations of group interaction. The highest level of agreement found by Bales and Hare (1965) in their review of 21 studies of group research using the IPA method was 22 per cent; this level of agreement being found in a student discussion group. Cline claimed that during difficult tasks and stressful situations, members of the group were more inclined to pursue relationship goals, supporting each other rather than dealing with the problem and enquiring about the risks involved. Pressure to agree may be so strong that group members may continue to agree blandly while unwittingly consenting to their own destruction. Such attributes are associated with groupthink.

Groupthink occurs when members of a group do not agree with statements that are made, although they do not make their view known to others, which results in the group members believing agreement is reached. Cline suggested a few ways of avoiding 'groupthink'; these included asking questions, noting an absence of disagreement (which serves as a warning to group members to reassess alternatives) and being aware that the risk of illusory agreement heightens as external stress increases. Hartley's (1997) review of communication research also points out that a seemingly unanimous agreement by the group may disguise a silent minority.

Meyers and Brasher's (1999) review of group communication research suggests that there has been too much focus on task-based attributes of interaction and calls for investigations on the influence of emotion-based communication, for example aggressiveness and protests and how they affect group interaction. The Bales' (1950, 1970) IPA tool has the capability to collect emotional data and the potential to investigate such issues.

## Defensiveness and barriers

The seminal work of Gibb (1961) is worthy of discussion at this point. Over a period of eight years, Gibb recorded discussions and conducted a series of experimental and field studies designed to investigate the arousal and maintenance of defensive behaviour in small groups. Defensiveness can be defined as the behaviour of an individual when he or she perceives threat in the group. A person who perceives threat may communicate in a guarded or attacking way. Defensiveness of this nature will manifest behaviour patterns that are either consciously or subconsciously recognisable to other parties. The inner feelings of defensiveness create outwardly defensive postures. If such actions take place without question an increasingly circular destructive response may occur. Defensive signals are said to distort the message. When a receiver attempts to understand a communicated message they also extend their efforts towards understanding the motives behind defensiveness, attempting to understand why someone is behaving in this manner. Excessive defensive and aggressive behaviour distracts those engaged in the interaction away from rational discussion. Defensive arousal prevents the listener from concentrating upon the message. The defensive behaviour increasingly distorts as the circular defensive behaviour continues. The conversation moves from the subject matter to defensive action and reciprocal attacks. Gibb (1961) suggests that the defensive and supportive communication can be accommodated without changing the content of the statement. For example, enquiring for, rather than demanding, information reduces defensiveness and aggressive responses. McCann (1993) suggests that good communicators develop flexible interaction techniques, enabling a greater appreciation of the other's perspective.

### Help-seeking and question asking

Help-seeking behaviour implies incompetence and dependence (Lee 1997), and many professionals are reluctant to ask questions for fear of being perceived in such a manner. Furthermore, most people are not very good at asking questions (Ellis and Fisher 1994). Hawkins and Power (1999) summarised the findings of previous research using the IPA method noting that 6–7 per cent of a group's total acts tended to be questions. However, using an adapted version of IPA, which focused more on questions than other communication acts, Hawkins and Power found that question asking constituted 16 per cent of the total turns taken by groups of undergraduate students. Hawkins and Power also noted that the difference in research methods could have resulted in the increased amount of questions. However, the higher level of question asking may also be due to the context and lack of status difference between the students.

Lee's laboratory experiments, which investigated the number of times a person sought help and asked questions, found that it is more likely that high status professionals will avoid situations where they need further information, in order to avoid asking questions and to defend their status. Lee noted that when teams were formed of high status members the number of questions asked reduced. By undertaking a field study of a hospital that had recently installed a new computer system, Lee confirmed that the results also apply to organisations. All of the nurses and physicians had knowledge gaps about how to use the computer systems and they needed help from either the specialist assistant, who was provided to help the staff operate the system, or those more familiar with the system. Using surveillance equipment Lee was able to monitor how the professionals sought help and who they approached. The results showed that the participants were less inclined to ask for help from higher status colleagues, and higher status colleagues were less inclined to seek assistance. In both cases, advice was often sought in informal environments during conversations. Other research has also found that serious and costly errors have been made in multidisciplinary projects which could have been prevented by seeking expert help that was available at the time, for example Capers and Lipton's (1993) candid research into the behaviours of engineers involved in the development of the Hubble Space Telescope. The engineers were found to avoid interaction with specialists employed to provide expert optical advice. The engineers' behaviour showed that they did not want faults to be seen by others and wanted to resolve problems on their own, even though they failed to resolve the faults.

In Mabry's (1989) laboratory research involving 27 female and 18 male students undertaking unstructured tasks in 4 groups, the males asked other males less questions compared with the females. Although asking questions is the single most effective way to extract ideas and information (Ellis and Fisher 1994), these studies suggest that where professionals do not understand a situation they may be reluctant to ask for help. Help-seeking behaviours are fundamentally inter-personal; one person seeks assistance from another (Lee 1997). Seeking help from others often occurs simultaneously with seeking information and feedback (Morrison 1993). Individuals are more likely to seek help from equal status peers (Morrison 1993; Lee 1997) and others who have helped them earlier; co-operative patterns are reciprocal (Patchen 1993). The use of question asking and seeking information within construction meetings requires further investigation; however, the method used to investigate interaction must be able to differen-tiate between interaction that is used to seek information and questions that may expose another person's weakness or threaten another. Some questions can be perceived as an accusation (e.g. 'who did that?') and can result in defensive arousal (Gibb 1961: 143), although Gibb notes that questions without overtones or negative emotion do not ordinarily result in defensive behaviour. Questions can be used to gain further information or understanding, seek help, accuse or threaten others. The Bales' IPA system classifies information seeking into three categories: these include requests for information, opinion and direction. Where questions have emotional overtones they would be classed into one of the socio-emotional categories. It is important to see how much effort is placed on gathering information from others and whether there is a difference in the level of information requests between those members who are more and less effective.

### *Functional and dysfunctional conflict*

Conflict has been found to develop in multidisciplinary building design teams as the group members discover their team objectives and then attempt to enforce them on others (Wallace 1987). This is true of competitive and collaborative arrangements. Situations change and evolve over the time frame of projects and it is impossible to predict all eventualities: it is inevitable that conflict will occur, for example in an attempt to avoid rework. Conflict can be natural, functional and constructive or unnatural, dysfunctional and destructive (Huseman *et al.* 1977; Gardiner and Simmons 1992). *Natural conflict* is described as the intended or actual consequence of encounter resulting in stronger participants benefiting from the clash. This is inevitable and thus plans to deal with it can be made in advance. *Unnatural conflict* is where a participant enters into the encounter intending the destruction or disablement of the other, usually with the intention of making a financial or personal gain. This is quite a well-known strategy of less scrupulous contractors looking to increase their profit margins on a project. One reason why parties have a strong desire to work with those who have a good reputation or with others where they have previously formed successful relationship is to avoid the commercially aggressive or 'claims conscious' contractor. Care should be taken when selecting the project participants. Further research is necessary to not only recognise the aggressive 'win at all costs' party but to develop strategies that can deal with them.

There is enormous disagreement over the effects of conflict on the group's social system and communication (Pondy 1967; Folger and Poole 1984; Ellis and

Fisher 1994). In Farmer and Roth's (1998) study of conflict using content analysis of audio recordings of 29 different group meetings from employee work groups, student groups, medical school teams, conservation committees, environmental planning committees, physicians and funding consortia, conflict emerged in all situations. Furthermore, conflict was found to be more prominent in the smaller groups. While different groups handled conflict differently, many of the group members accommodated the conflict behaviour, failing to attempt to satisfy their own concerns. Farmer and Roth concluded that such action may result in process losses (ideas and information being suppressed); as participants fail to argue for their concerns the synergies from different inputs and ideas may be lost. Whereas, when parties are assertive and co-operative, collaborating to satisfy the concerns of all parties, delving deeper into issues and exploring disagreement, the process gains from the aggregated group inputs. Conflict is important to expose and explore risks and alternatives. Equally, conflict needs to be managed so that it does not suppress information or become personal and dysfunctional and damage relationships.

A number of reports suggest that conflict has advantages as well as disadvantages, although few go so far as to say that conflict should be actively encouraged. Loosemore *et al.* (2000) present a strong view on the importance of conflict within the construction industry. Loosemore *et al.* claim that indiscriminate reductions in conflict incurs an opportunity cost for construction clients. The research, which focused on the construction phase and conflict between the contractor and architect, and contractor and sub-contractor, found that the contractors' attitudes and the social structure of the construction system were receptive to functional conflict, although not as strongly as Loosemore originally thought. Rahim's (1983) independent scales for measuring conflict-handling styles were distributed to 300 site managers involved in a broad range of commercial, domestic and industrial developments. The findings suggest that most conflicts were managed by exploring alternative solutions, different perspectives and encouraging all participants to engage in discussions and co-operate. Such behaviours were believed to be most likely to result in win–win solutions. However, a considerable proportion of the site managers' conflict-handling style was not considered to offer such positive benefits. Too much emphasis on compromising and obliging by the site manager restricted the potential development of mutually beneficial solutions. While site managers are often considered to be uncooperative and lacking concern for others, the style of conflict management, which was least used, was the dominating style, which places concern for self before others. Loosemore *et al.* concluded that teams should foster more integrative styles of conflict management, balancing both concerns for others and self interests. While such approaches are encouraged, little information is presented on how members should interact to engage in and manage conflict. Before attempting to manage conflict, consideration must be given to the group's regulatory system, how conflicts emerge, how they are controlled and how task objectives can be achieved while maintaining relationships.

Early laboratory studies of small groups by Heinicke and Bales (1953) found that higher status members in high consensus groups were more involved in socio-emotional conflict than others (status was based on observers' and subjects' perceptions of influence and leadership). The nature of the task has also been found to affect levels of conflict. In observation of interaction in 44 student groups completing structured (ranking tasks) and unstructured tasks (case studies), Mabry (1985, 1989) found that there was a higher rate of disagreement in structured tasks. In unstructured tasks, group members had to first understand the problem before they could disagree on issues. Using the Bales' IPA, the levels of disagreement were found to be related to high levels of information given and requested, and lower levels of opinions. The differences in levels of disagreement were relatively small, although significant. It is rather difficult to conclude whether such findings are a result of what were considered unstructured and structured tasks or the idiosyncrasies that result from different tasks.

Benefits gained from conflict include increased understanding of issues and opinions; greater cohesiveness and motivation (Ellis and Fisher 1994). Through argument, challenge and conflict, group members are forced to see that others hold strong and defensible positions. Attacks on proposals will mean that members have to defend and justify their ideas, beliefs and suggestions. When members disagree and explore why they disagree, they expose key issues and points of misunderstanding. If conflict is managed issues are explored in depth. Groups that experience tension and conflict and then work through these experiences often feel closer and stronger; Loosemore's (1996a) research found that this sometimes occurs following a crisis.

Disadvantages that may be experienced include decreased group cohesion, weakening of relationships, ill feelings and destruction of the group (Ellis and Fisher 1994). If conflict goes on too long and is not resolved, it will decrease cohesiveness within the group; conflict between people can be distasteful and personalised, having little relevance to the task or problem. Most people do not like to be criticised and all conflict has a negative socio-emotional impact, which must be recognised. When conflict takes place it is important that the impact on individuals is tempered with some form of positive socio-emotional support. The support does not need to come from the person who initiated the criticism, but the group's relationships must be managed. Groups that do not work through conflict and repair relationships will fall apart. Conflicts that turn into disputes incur major resource costs, can disable a company, and often results in significant expense for both companies. The negative perceptions that develop from public dispute often serve to damage the winning and losing parties. Emotions that become embroiled and inflated during disputes normally mean that the issue that was at the heart of the dispute becomes of little concern. Thus, public disputes normally result in process losses rather than gains.

It is inevitable that conflict will occur in groups (Loosemore 1996a; Farmer and Roth 1998). Bales (1950, 1953) recognised that individual goals associated with performing tasks and maintaining relationships can sometimes be in conflict with other group members' goals. Groups and individuals must confront conflict to balance group and personal goals, attempting to pull the group back to equilibrium (Bales 1953), where a balance is maintained between the task and relationships. It is frequently argued that functional groups need to maintain relational as well as task-based communication if they are to perform effectively, especially during conflict (Bales 1950).

### Participation: Conflict and support

Individual levels of participation in a group have been linked to group conflict and co-operation (Ackoff 1966; Hancock and Sorrentino 1980; Wallace 1987). Using content analysis to identify disagreement, tension and non-conformist behaviour and measuring the number of times members participate in group interaction, previous research has noted that as interaction increases so does the occurrence of conflict. Wallace's (1987) longitudinal and cross-sectional study of design team interaction found that arguments between construction design participants continue to re-emerge over a period of time; thus, as opportunities to interact increase, the chances of conflict reoccurring also increases. However, Hancock and Sorrentino's (1980) study of group interaction found that where a group member had previously received support from other group members he or she is more likely to participate in a conformist manner on that subject during future group interactions rather than engaging in conflict.

Lawson (1970, 1972) suggests that greater amounts of information increase levels of conflict. High levels of information and conflict are often experienced towards the end of projects. Hancock and Sorrentino (1980) found that an upsurge in conflict may be self-propagating as a result of conflict being carried forward and expanded. People tend to be more inclined to express their emotions when they are more familiar with each other: thus the longer groups have been together the more likely negative emotions will be shown. Established groups may not necessarily experience less conflict than newer groups, but they tend to be better equipped for dealing with it.

During a group's development a more defined structure of interaction evolves through the group's regulatory procedures (Wallace 1987). Through experience, group members learn to expect conflict in certain areas between certain members (Lieberman *et al.* 1969; Schutz 1973; Hancock and Sorrentino 1980), and also gain support from other members. Wallace (1987) supports this view suggesting that, as the group's socio-emotional awareness develops, the members anticipate where potential conflict could manifest and use supportive reinforcement interactions to overcome conflict between members. This allows participants to engage in task-related elements and control discussions with socio-emotional interludes (Bales 1950; Wallace 1987).

Deutsch (1949) suggests that higher levels of conflict in the later stages of a group's life tend to produce greater diversification in group members' participation. However, Wallace (1987) and Loosemore (1996a) found that competition and conflict between group members produce co-operation between others.

Co-operation between subgroups was characterised by a decrease in conflict and an increase in supportive contributions.

A certain amount of conflict within any organisation is inevitable and the existence of communication problems will make the management of conflict difficult. The management of conflict within organisations needs to concern itself with the reduction and eradication of dysfunctional conflict and the manifestation and management of functional conflict, which will help discussions remain creative and useful. When there is more than one organisation affected by conflict, such as in a project environment, each organisation will attempt to secure their own goals before addressing those of the temporary construction organisation (Loosemore 1996a). Loosemore observed evasive and defensive behaviour during a crisis. Through interviews and diary reports he was able to deduce that such behaviour was aimed at reducing or minimising increased commitment of an organisation's resources to the construction project.

Klob (1992) believes that a supportive group climate should be developed so that when conflicts emerge, as they inevitably will, the group is able to repair emotional damage and continue with the group task. Conflicts should not be avoided but managed (Klob 1992). Bales (1950, 1953) argued that relational damage caused by critical discussion is repaired by positive emotional expression, such as showing support, passive agreement etc.

## Group performance and outcomes

Clampitt and Downs (1993) note that intuitive links between communication and productivity make sense. They also cited a number of surveys that showed perceptions on this relationship were strong. However, their research, based on the interviews of 175 employees from two companies, which were analysed using content analysis, found that perceptions of productivity were diverse and the link between communication and performance was more complex than previously assumed.

There are issues that need to be considered when attempting to measure the effectiveness of group performance. Critics of theories that are linked to outcomes argue that assessing the performance and outcomes of group interaction is very complex and the conclusions are far from straightforward (Billingsley 1993; Stohl and Holmes 1993). Valid criteria for judging the effectiveness of real world decisions are difficult to define and often conflict; what might appear to be a successful short-term decision may result in long-term problems, and vice versa (Poole 1999). Success is a subjective factor and what one party perceives as success another can view as failure. The successful completion of most construction projects normally results in some being happy with the end product while others dislike the building or could have been against the project from the outset.

A humanist perspective on success focuses on satisfaction of parties during group activity or contentment with group behaviour, project-oriented perspectives focus on achieving tasks and goals. In short-term projects, sustaining relationships between members may only be important to achieving the goal. The success of long-term relationships and repeat business is dependent on satisfying both relationship (humanistic needs) and project needs, thus there is a balance between social and task dimensions of success.

Observers' values, background and perspective will affect their assessment of group performance. Poole *et al.* (1999) found that when different groups of people evaluate group performance, differences are often found. External evaluators have been found to evaluate group performance differently, taking a more negative view of group performance than group members. Both observers and group members are said to contain bias, and one method of reducing the bias is for group members, external observers and specialists to evaluate the same performance, producing a triangulated view (Poole *et al.* 1999).

Hackman and Morris (1975) formed and studied 108 groups undertaking 4 intellectual tasks for a period of 15 minutes for each task. The groups' written solution output was measured against action, length, originality, quality of presentation, optimism and issue involvement. The group interaction was coded and analysed by a 16-category coding system. Despite their efforts, no consistent relationship was found between the groups' interaction variables and performance. The conclusion reached was that the apparent complexity of the relationship makes it difficult to develop simple explanations of how interaction affects performance. Jarboe (1988) using IPA (as well as some unique categories) examined the effect of input, process and procedures on the output (quality of solution). Although Jarboe found no relationship between procedures and quality of solutions, interaction categories that were found to have an effect on the quality of solution were communication that involved suggesting solutions, offering direction and generally showing autonomy. Consistent with previous findings, direct communication has been shown to have the potential to influence group behaviour.

Many of these studies, such as Jarboe's, are based on member satisfaction or broad perceptions of performance and are quite subjective. Because of the values people hold, performance data should be clearly defined and measures should be assessed against the defined criteria, reducing the subjective assessment. Such objective measures do limit the scope of studies, but are clear on what they are measuring and what criteria are used to assess 'success'. Data on satisfaction are often based on self-reports; however, people are not always aware of the causes of their satisfaction. Although it is difficult to draw causal relationship between group interaction and output, it may be possible to identify the practices of groups that are associated with specified performance; a characteristic that is of interest to design and construction teams.

## Summary

Conflict is an essential part of organisational activity. Previous research claimed that arguments and disagreements might be avoided so that members can pursue relationship goals. The balance between disagreement and agreement in a project environment may be difficult. During group meetings, issues have to be discussed with sufficient rigor to produce the optimal solution, but relationships must be sufficiently maintained so that members are able to continue to operate as a team. Failure of participants to make effective contributions to group discussions will reduce the group's decision-making ability. It is inevitable that some members of the construction team will be reluctant communicators. Similarly more active communicators may restrict their interaction to task-based discussions in an attempt to avoid emotional exchanges. The temporariness and fragility of the construction team may also restrict the nature of interaction. During early stages of group interaction, the participants confine their communication to task-based messages; emotional exchanges emerge as members become more familiar with each other. Where specialist knowledge is required and the professional does not make a full contribution, the ability of the group to make an informed decision is reduced. The characteristics of group interaction will be affected by the development of the group norms. Group norms may constrain or encourage participation and the use of certain communication acts. In particular the norms associated with construction meetings deserve more attention.

The way in which a group develops interaction norms will affect the group's ability to successfully discuss tasks, evaluate proposals and maintain relationships. The findings of earlier research suggest that question asking, giving directions, and use of emotional interaction can affect group behaviour. The development of communication mechanisms that inhibit and enhance the exchange of information within the group is fundamental to the team's performance. Although it may be difficult to show causal relationships, it is possible to examine the group interaction patterns associated with successful outcomes and subsequently analyse the communication traits of those individuals considered to be more successful than their peers.

# 5 Collecting communication data from construction projects

The PhDs reviewed earlier help to highlight some of the principal difficulties of gaining access to sensitive business environments and extracting appropriate scientific data from the field. Indeed, one of the reasons why so little research has been published on communication in construction teams may be related to the methodological challenges inherent in trying to gain permission to access and record interactions with dynamic and sensitive business environments (Emmitt and Gorse 2003). Gaining access to business settings is difficult and often exacerbated by the social scientist's desire for rigour and use of a method capable of capturing the data in its richest (most real) form (Bryman 1988). Negotiation with senior members of organisations is necessary to reach agreement on the research method to be used, the amount and type of observation methods, the manner in which data is recorded and the dissemination strategy to be employed, before data collection can commence. Negotiation is dependent on the gatekeeper's belief that helping the researcher is a good gesture, or the research has some benefit either to the company and/or benefits the wider scientific field. Gatekeepers also need to be convinced of the researcher's integrity and of the value of the research before they decide to cooperate (Bell 1999). Once this has been granted the researcher is then often faced with the added difficulty of getting the approval from the individuals who are subject to the research.

Collecting data during construction negotiations can be problematic because information is seen as a confidential source of power (Loosemore 1998b). During a construction project actors frequently discuss activities and make decisions that ultimately lead to an allocation, or redistribution, of resources. The continued success of an organisation is dependent on its effective allocation and use of resources; subsequently, the financial pressures on the negotiators may be high. To ensure financial success and achieve their desired outcome, actors may move through informal and formal discussions using whatever communication skills and powers of influence they have. At times the discussion may become emotional, argumentative, involve brinkmanship and result in behaviours that may become somewhat uncontrolled and irrational, with the use of dialogue that some may consider inappropriate. The fear is that participants may unwittingly discuss confidential issues or make statements that could threaten their reputation, if made public. Not surprisingly participants may be very wary about their conversations being recorded or even witnessed, especially if discussions are perceived to be of a highly confidential nature. Non-participant observation by a third party (the researcher) adds no immediate value to the participants' business processes and the perceived threat of someone unknown (and hence not trusted) entering their private space may outweigh any benefits associated with the research. Even when researchers have been given permission to record communication interaction it is not uncommon for participants to ask researchers to switch their devices off, not to write anything down or even to leave the room if discussions are perceived to be too sensitive by one or more participants.

Data collection methods must be appropriate to the context, being both capable of extracting relevant data and acceptable to those being observed. For example, capturing interaction using video and audio recording can provide a rich source of data, but professionals are often reluctant to let such equipment be used in the business setting. Our experience is that less resistance is encountered when researchers use written records to collect data from the field (Gorse 2002; Gorse and Emmitt 2005a,b). Although this may reduce the richness of the data collected there may be no other way of extracting data from the situation and some compromises may be difficult to avoid.

## Observing interaction

Communicators simultaneously send and receive multiple signals, which are stimulated and responded to by different levels of intellectual and emotional processing (Kreps 1989). The communicators are consciously aware of much of the information exchanged verbally and through non-verbal interaction; however, the sender and receiver may not be consciously aware of powerful signals that they send and receive while the subconscious mind still processes and reacts to those same signals. In small groups and even in one-to-one meetings, messages are sometimes cryptic with more meaning to certain individuals; participants who do not have the same background knowledge have little chance of recognising the relevance and significance of such messages. The nature and dynamics of interaction within groups can be considered to be both observable and hidden (Hugill 1998). Signals are sent at a subconscious and conscious level, although observations of communication are limited to aspects that are consciously processed. Observers can record interpersonal communication acts transmitted through expressions, sounds, actions and reactions. Thoughts and beliefs can only be accessed by retrospective explanations, records or accounts supplied by the person in whom the thoughts manifest.

Messages processed by the subconscious mind can be hidden from those interacting, yet may still be recognised and observed by a third party. Even though actors are unaware of the signals they process subconsciously, others may recognise and respond to them. The Johari window (Luft 1984) is a useful concept to consider when thinking about the hidden and disclosed aspects of human behaviour (Gorse and Emmitt 2003 and Fryer *et al.* 2004 provide detailed explanations of the theory). Simply put, the Johari window divides aspects of human behaviour into four areas:

- *Hidden information*: Information, beliefs and thoughts that are known only to our self.
- *Open information*: Information known to our self and openly disclosed to others.
- *Information known only to others*: Behaviour and information that is recognised and known to others, but not consciously recognised by our self.
- *Information known neither by self nor by others*: Aspects of behaviour that are totally hidden from both participants and recipients (although the behaviour or information exists, those engaging in discussions or involved in the relationship are not consciously aware of it).

A variety of research tools can be used to investigate these areas. Diaries, interviews and retrospective accounts can be used to delve into thoughts and beliefs that were previously private and hidden. Information that is open can be accessed using a range of observation and inquiring techniques. Third party observation and specialist analysis of audio and video data are particularly good at identifying behaviour that is neither recognised by the self nor by others participating in the group. It is inevitable that the research team will fail to record all thoughts, signals and behaviours given the nature of communication. It is to be expected that the person, participants or the researcher, or all of these parties will not be consciously aware of all of the signals exchanged, yet such information could be processed subconsciously. It is within these limitations that all aspects of human research are conducted.

## Quantitative observation with qualitative explanation

Social scientists have developed and tested a number of methods that have been used to study different social contexts (Clark 1991; Frey 1994). The researchers of construction communication, using the tools of the social scientist, have identified strengths, weaknesses and degrees of applicability that various research methods have when applied to the study of construction processes (e.g. Gameson 1992; Loosemore 1993, 1996a,b, 1998b; Simister 1995; Hugill 1998, 2001; Gorse 2002; Abadi 2005).

### *Reliable and consistent measurement of communication*

Research may seek to classify and categorise communication acts, enabling the development of models based on a particular classification system. A difficulty associated with studies of this nature is that they must transform communication that is continuous, intermingled, overlapping and abstract into observable phenomena. For classifications to be reliable, observations must also be categorised with significant degrees of confidence (Clark 1991). Aspects of communication must be described so that they are not abstract, but observable acts and occurrences. Thus, the study of communication is not easy and care must be taken to select or develop an appropriate methodology that is reliable, consistent and which produces valid results.

Hargie *et al.* (1999) argued that communication can be characterised in a scientific manner because it is a process that is open to measurement, analysis and improvement. The translation of behaviour into communication acts allows the data not only to be segregated into categories, but also to be aggregated together providing opportunity to analyse groups of categories. Once captured and classified the interaction data of an individual, the group or multiple groups can be compared and analysed. Use of specific categories may be more or less prominent during certain types of encounter or between specific members of a group. Thus, systems such as Bales' IPA enable differences between individuals to be identified across 12 categories of behaviour. Classification and quantification of data enables trends and patterns of behaviour to become apparent that would otherwise be missed. Multiple comparisons of interpersonal exchanges between different actors are easily produced with the aid of spread-sheets or statistical processing software. Once trends are identified it is often useful to investigate the specific events with more qualitative approaches to determine the exact nature of behaviour that is responsible for the pattern of interaction.

Research on communication in organisational settings is considered by some researchers as too complex or inappropriate to model using quantitative methods (Cassell and Symon 1994). Use of quantitative methods alone offers a limited perspective since the experimental unit associated with a temporary multidisciplinary organisation is not one that can be totally controlled. The research unit being observed develops, changes and responds in different ways depending on how the participants behave and act out their roles. There is a danger, particularly with statistical methods, of becoming too focused on the intricacies of measuring, and thereby focusing attention on the classification of interaction instead of observing what is actually happening (Cassell and Symon 1994). Classification systems by their nature are very tightly defined to enable researchers to fix the observations to specific acts of communication. Actions and events that fall outside the prescribed categories will either be missed or may be recorded incorrectly, introducing inconsistencies to the recorded data. This problem is more than offset by the advantages afforded by a system of recording with predefined categories, allowing the research to be focused and helping the researcher to collect data in a challenging environment such as construction. Rather than attempting to collect everything, the narrow categories allow specific events to be recorded, quantified and analysed allowing trends to be identified that would be missed if a broader more qualitative system were used.

For every advantage that a quantitative observational system possesses in terms of effectively capturing some aspect of reality, there will be corresponding losses in terms of the abstraction that such systems invariably demand (Halpin 1990). For example, in the Bales' IPA system the coding of the communication act ignores information relating to its significance, intensity and salience. The system does not capture the motives behind communication nor does it recognise the participants' feelings and beliefs during interaction. Many scholars also stress the need to add a qualitative dimension to the observation of communication. Some feel that there is a need for the researcher's qualitative comments to support the quantitative observations (Hirokawa and Poole 1996; Cappella 1997; Philipsen and Albrecht 1997; Frey 1999). Even those whose work is based mainly on quantitative analysis, such as Bales' (1950, 1953), call for qualitative comments to explain the interaction. A third party's interpretation of the acts and reactions can be useful. The use of reflective commentary provides an additional perspective to the theory and an extra dimension to the research enabling greater understanding of how the interaction develops (Cappella 1997; Philipsen and Albrecht 1997). Observations of group communication behaviour that fall outside classification systems may be helpful in understanding how communication acts are used, helping to contextualise the data. Commentary on the behaviour of groups during interaction sequences will provide information on the immediate effects, behaviour and reactions of members' communication. Explanation of interaction will help differentiate behaviours that would belong to the same category. Expressive information could be captured, for example intonation, emotional expression and intensity with which the communication act was used, which would often be outside the scope of the categorisation system. Such observations are directly linked to the group communication behaviour observed and help to differentiate the significance of communication acts that may be captured within the same communication category.

### Ethnomethodology

Rather than quantifying and classifying interaction, Hugill (2001) adopted an ethnographic approach making use of conversation analysis to study group interaction during construction progress meetings. Ethnomethodology aims to develop patterns of thought about that which is observed (Silverman 2001; Sanders *et al.* 2003). The method provides insights into what, how and why the subjects' patterns of behaviour occur (Fellows and Liu 1999). Features of ethnographic research include (Silverman 2001; Abadi 2005):

- Emphasis is on the nature and understanding of the social phenomena rather than attempting to test any particular hypothesis. The method aims to develop an understanding of reality.
- The method uses original, non-coded, unstructured data. As far as possible the data should be captured in its raw form. Even though they have their own limitations, raw transcripts and audio and video data are a preferred method of capturing the experience, but these support the experience. It is important that the researcher is present at the situation being studied.

- Due to the richness of the qualitative data, investigations can be limited to a few case studies; on occasion the data may be limited to one case study or part of a discussion.
- Analysis of the data is based on interpretations of meanings and functions. The interpretations must be explicit. Verbal descriptions and explanations are the product of ethnographic research.

Ethnography aims to describe the social experience of the group being studied from the participants' viewpoint. The researcher is usually described as a participant observer, their presence makes them part of the process (Jankowicz 2005). The study attempts to provide an account of what the participants notice as meaningful actions, events and statements. To be true to the nature of the ethnographic study the data should be described in the participant's own language. The emphasis is placed on those aspects that the participants wish to emphasise. Because the researcher wants to see and experience the world through the eyes of the participants and not be placed under the influence of external sources, the researcher is discouraged from undertaking the literature review prior to the main study (Jankowicz 2005). However, attempting to remove previous assumptions, when researchers already have an interest in the area being studied obviously proves to be practically impossible, especially for those who research specific areas for prolonged periods of time.

The 'Hermeneutic interpretation technique' can be used to identify and interpret the meanings expressed in the data. Parts of the method are also useful for interpreting and analysing research regardless of whether ethnomethodology is used. The technique relies on the following steps (Jankowicz 2005):

- Attempt to understand the meanings expressed by the participants either in the text, objects or events that manifest.
- Identify trends and sub-themes.
- Identify clusters of themes which occur across text or events.
- Compare and triangulate with other data.
- Check the validity and reliability of interpretation with other people.
- Ensure that the themes are placed within the correct original context.
- Sample key documents, being systematic and preparing a case report – which will represent a conclusion.

Ethnomethodolgy and the subsequent analysis process are concerned with the contextual sensitivity of language with a focus on dialogue as a vehicle for social action. Investigations using conversation analysis can only be pursued through intensive qualitative analysis of interaction events, because conversation data proves quite resistant to treatment in terms of normal sociolinguistic variables and quantification (Drew and Heritage 1992). Transcripts, video or audio recordings of interaction are required to provide the data necessary for conversation analysis. The detail of analysis required can often restrict the amount of data analysed, for example Hugill recorded 30 hours of construction team discussion but analysed less than 1 hour of data. Although the use of conversation analysis does not lend itself to studies involving multiple observations of different projects, qualitative observations are considered to be important when examining group interaction. Ethnomethodology is not suited to the study of multiple groups over long periods of time; however, it is capable of developing meaning and understanding to specific episodes of communication where other quantitative tools would not. Because the data extracted is that of the subjects' verbal utterances, confidentiality and anonymity can be a problem in business settings. It should also be noted that transcripts from group discussions rarely capture all of the utterance made.

### Transcripts and video footage

The act of transcribing data is not always as easy or as precise as one would necessarily think. In a study conducted by Gorse and Emmitt (2005a,b) a group of 18 research students were given the task of observing and analysing group behaviour using different data collection methods and research tools. All students took part in one of three 30-minutes group discussions, which were recorded on video tape. Each student was given a video copy of the group discussion and asked to transcribe the data and then analyse it. Most research text assumes the transcription of audio data to be relatively straightforward and offer little or no guidance on how transcripts should be prepared so that they produce consistent results. At the outset the students were asked to reflect on all of the research methods and tools used to record the interaction. The results provide some interesting perspectives on the preparation and use of transcripts and use of video data (Tables 5.1 and 5.2). It is important to note the degree of variation that can be found in the act of transcribing. Although transcripts can be prepared from audio records, we should recognise that the consistency of the transcripts may vary according to the abilities of the transcribers. To improve consistency, transcribers should be given guidance on the level or detail expected to be extracted, especially in the area of overlapping conversations, muffled and whispered comments. Clearly there are many advantages to the use of video and audio footage, but it may be difficult to use in professional environments.

### Different perspectives

Different perspectives are gained through the use of different methodologies (Seymour and Hill 1993) and when triangulated they can help develop better understanding than a single method. Combined use of quantitative and qualitative methods increases the detail of the information collected thereby improving the overall methodology and hence reducing some of the research limitations (Fielding and Fielding 1986; Azam *et al.* 1998). Quantitative data is useful for summarising and analysing large sets of data whereas qualitative approaches can enquire into specific instances and events. Thus, qualitative analysis of interaction behaviour data could help to aid interpretation of quantitative data.

*Table 5.1* Reflections on the processing and use of transcription

| Strengths | Limitations |
| --- | --- |
| Provides a general overview of the meeting. | Transcription is time-consuming. |
| Audit trail of all of the verbal messages sent. Every sentence and word recorded. | The distillation of video data into words varies. For example, the time taken for one research to transcribe the data was $5^1/_2$ hours, another $9^1/_2$ hours and further researcher 3 days to transcribe 30 minutes footage. One researcher employed a professional audio typist, but still found that it took hours to turn the type into a proper transcript. |
| The transcribing process allowed for a better understanding of what was said, helped to understand some group dynamics; this was considered an advantage during the later stages of the analysis. | |
| The transcript and video can also be used with other data, such as Bales' IPA, self-perception profiles and Belbin self-perceptions. | Some transcribers record more than others. It is a difficult and confusing task to record transcripts, and it is impossible to track every nuance of the conversations. |
| Useful to focus in on the interaction trait, to analyse in depth. | It is difficult to transcribe muffled speech, people talking over each other, and attempted interruption. |
| Detailed nuances of the video would be very difficult to follow without the support of the transcript. | |
| Once the classification data identifies certain tendencies, these can be investigated in greater detail using the transcripts and videos. | Transcripts ignore how the message was sent, body language, eye contact, intonation, tone, emotion and humour. |
| Time frames should be recorded on all data so that they can be easily compared and cross-referenced to other data. | Unless the research is strongly tied to the transcriptions, there may be a limited need for a transcript. Considering the time it takes to produce the transcriptions, some thought should be given as to whether this is a worthwhile exercise. |
| After repeatedly watching and transcribing a video footage, a general distinction of the contribution of each member can be made and evaluated, for example the frequency of each speaker, arguments and other occurrences. | On its own the transcript does not really constitute a systematic study. |
| Transcriptions provide a qualitative piece of work, which is a useful reference document. It is sometimes hard to locate the piece of transcription that you are looking for. It is important that appropriate coding is used alongside the transcription. | Transcripts fail to record whom the message was sent to. |
| | When analysing the data, looking for specific quotes or searching through the data can be a painstaking procedure. |
| | Some observers add their own observations to the transcript, introducing an element of subject interpretation into the raw data. |
| Video data and transcripts are useful for those who are less familiar with the language – foreign researchers, observers from different industries and professional backgrounds. | Transcripts compiled by different researchers are often slightly different. |
| | Transcripts should be used as a secondary tool, in combination with the video to see how the 'live' communication took place. |
| | It is easy to create the wrong picture by just looking at the transcript or video data. |

*Table 5.2* Video data and observations/reflection of video data

| Strengths | Weaknesses |
|---|---|
| Events and situations can be observed and reflected on. | Can be difficult to interpret any underlying intentions, may be considered wrong to infer intentions. Other methods should be used to identify participants' intentions during a specific sequence of event. Participants could be asked to review the video and asked what their intentions, thoughts, beliefs etc. were during the specific episode of events. |
| Provides initial understanding of the group interaction, good base to work from. | |
| Accurate account of discussion. | |
| Data set in context, some surrounding information available. | |
| Can judge positive and negative reaction, often missed in other data. | |
| Body language, voice tone, intonation and emotion can be examined. Emotional context can change literal meanings of words considerably, which is missed in transcripts. | While the data is rich and real, it is still difficult to capture every communication act, participants talk over each other and interrupt. |
| Disjointed conversations make much more sense and gain a congruent reaction. Transcripts of such event are meaningless. | Some utterances and statements may not make sense. |
| Recordings are essential to recall what happened. | Video observation is time-consuming. |
| Without video data other analysis can be inaccurate. | Emotion and body language can be observed, but it can be very difficult to transfer non-verbal observations into the written form. |
| Useful for cross-examining perceptions, whether individuals do what they say do. | |
| Participant, non-participant and external observation can be used to assess data. | Relying on video data without proper analysis can be too simplistic and subjective, lacks systematic rigor, no way of judging whether a group is typical or not. |
| Can be watched by many different people – obtaining multiple observations. Can be used for many different purposes. | |
| Requires little training, however, quality of observations dependent on researcher's training, experience and skill. | Subjects are aware of the camera and their behaviour may be affected. |
| Can be repeatedly reviewed to capture the subtle nuances of interaction. | How does an observer record what is going on when it may not be clear to the group? |
| Recollections of events, based on memory alone, are sometimes different from the video evidence. | |
| Issues can be examined in detail, researchers can look at what created these scenarios and how the group collectively and, or, the individual reacted. | Camera positions mean that behaviour is missed, multiple cameras may be required. |
| Allows relationships, dominance, blocking, conflict, leadership, seating arrangements etc. to be assessed. | Individuals may hide their interaction from the camera. |
| Facilitates the use of other analytical methods. | |
| Can be played at different speeds, this often identifies behaviours not apparent at normal speed. In the fast forward mode it is easy to notice the members who remain motionless and others who fidget or move. | |

We considered it important that the research presented here should build on existing theories, offering an empirical perspective. Similarly, to identify differences between groups we felt it was important to attempt to measure relationships between variables systematically and statistically using quantitative methods. In adopting such an approach it was recognised that the flexibility and scope of study would be constricted. Adding qualitative observations, such as reflective commentary, to aid interpretation of the quantitative data would help to reduce these limitations and enrich the data. We also wanted to collect interaction data from multiple meetings on different projects. With many meetings exceeding one-hour duration, producing vast quantities of data, the method selected needed to be capable of capturing the data in a consistent manner. Analysis would involve looking at trends across all of the data or that collected from a single project or specific participant. Comparison between projects and participants, as well as previously reported studies, was a further consideration. Classification and quantification of data using an established system would allow the use of descriptive and inferential statistics to be applied to the data as a whole and to individual data sets. Reliance on purely qualitative methods, such as ethnomethodology, restricts analysis to relatively small sets of data or particular episodes of interaction and requires data capture on video or audio recording devices.

## Clarifying, classifying and coding communication acts

The categorisation of communication is often used to allow quantities to be generated from observations of interaction; however, there are many different units of analysis that can be recorded (Burke 1974). The most basic units include:

- *Participation*: Participation identifies the contributors, those who are actively involved in the communication behaviour being observed (Stephen and Mishler 1952).
- *Act*: The communication act defines and classifies communication into discrete acts of verbal and non-verbal communication (Bales 1950).

A core assumption of many group communication theories is that communication is the observable phenomenon binding together the systemic entities of the group (Mabry 1999). Most communication models are based on observations of external factors or indicators of communication, such as the sending and receiving of verbal and written messages, facial expressions, emotions and body language, or reactions to these messages. Observation of overt external factors of communication, identifying who makes the communication act and whom the communication is specifically directed at, has been termed the 'surface meaning' of communication (Heinicke and Bales 1953). Surface meaning identifies the acts of communication that are most obvious to the observers, it does not attempt to identify motives, beliefs or cryptic messages.

Coding systems are used by communication researchers to group together related communication acts under a common heading. Different categories can be used to identify aspects of communication that can be observed, tabulated and compared. A coding system is simply a lens through which the researcher has chosen to view events (Bakeman and Gottman 1997). Bakeman and Gottman go on to say that a good coding system can be used to generate a better understanding of events, whereas an ill-conceived coding scheme has little use.

One of the most fundamental components of the communication research process is discovering and documenting the discrete action-based elements, the signals, gestures, units of interaction and patterns, and to specify the relationship obtaining between these elements (Duncan and Fiske 1977). To capture and accurately record the acts they need to be described with sufficient clarity and in detail so that others can recognise and observe the same act. The methods must be capable of being learned and administered by others to produce consistent and reliable results. Duncan and Fiske (1977) believe that observations of face-to-face interaction should be generated from a disciplined observation of categories of action, which require a narrow focus of attention, minimum levels of inference, discrete decisions, and moment-to-moment judgement. They also add that coders should avoid attribution of meaning and intent to the interaction. Contrary to this view, Bales (1950) claims that the researchers need to use, and cannot avoid using, their own intuition, considering the statements made before and after each observed communication act. Poole *et al.* (1999) discussed the difference in viewpoints and concluded that observers can take two stances when attempting to code information: either observer-privileged meaning or subject-privileged meaning. The observer-privileged meanings are attributes of understanding that are accessible to observers external to the specific group context, and subject-privileged meanings are the understandings that only the group's participants would have of the same incidents. Poole *et al.* (1999) noted that it is harder to capture the subject-privileged meaning than it is to the observer-privileged meaning. Capturing the subject-privileged meaning would involve asking the participants to discuss or record their actions and motives with regard to specific incidents of communication or group behaviour. The observer-privileged meaning does not understand the nature of relationships that exist within the group or the group knowledge; however, the external perspective has the advantage that it is not biased by relationships and motives that would exist within the group. Some systems such as Bales *et al.*'s (1979) SYMLOG and Fryer *et al.*'s SMOT (2004) analysis make use of third party observers (observer privileged) and participants (subject privileged) to capture as much data on the group interaction and experience as possible.

Poole *et al.* posit that the observer, when using the Bales' IPA system (what may be considered the observer-privileged position), takes the position of a receiver, rather than being blind to the group communication. They argue that such approaches allow bias when attributing meaning, bias being influenced by the observer's desire to support research hypotheses. However, the IPA protocol when applied correctly can be sufficiently rigorous to prevent bias towards hypotheses (Bales 1950) and it is not capable of capturing subject-privileged information (Gameson 1992). The system is blind to the purpose of the group, being objectivist and yet is still able to use observer-privileged meaning.

The Bales' system does not attempt to add something that is not there, it attempts to code the function of the communication act. Codification is not based on words alone, but our understanding of words when used in a context of interaction sequences supported by the reactions and emotional expressions of the group and its individual members (Bales 1950). It is important to recognise discrete interaction variables within the context of the larger interaction sequence. Discrete observations that fail to recognise how a gesture, comment or interaction was used can sometimes be misleading. Bales highlighted the importance of using our ability to understand language to identify the real meaning of words. For example, coding a sarcastic statement as an act of praise, if it is not meant to praise, would be incorrect. Researchers can easily fall into this trap especially if using transcripts alone to code communication. If the words were written on paper and read by a third party who was not present at the group meeting, their literal meaning may be attributed to a wrong category, for example coding an agreement as an agreement when the act was a sarcastic statement. To the observer of real interaction such attributes are much more obvious being supported by emotions and reactions of individuals and the group. If meanings are not attributed to words, or interaction, then communication is a collection of words without meaning. The distinction needs to be made between a system that attempts to understand the specialist nature of discussion and methods that attempt to capture the general intent of interaction. The IPA method classifies acts on the basis of simple emotions and task-based categories that can have positive and negative connotations. The system makes no attempt to capture specialist subject information.

The findings of a coding system are tied to the method used to capture the data (Poole *et al.* 1999). It follows that there will be some limitations involved with all coding systems, but more importantly there will be difficulties in comparing results obtained from different systems. The major advantage of observer rating, as used in Bales' IPA, is that once the researcher has been allowed into the environment where communication is taking place the ratings are easy to obtain (Clark 1991).

It may be possible to observe the subjects, taking ratings without their knowledge, in the natural environment, for example by using hidden recording devices or one-way mirrors or being present in the environment under the guise of a normal participant. Unfortunately, from a researcher's perspective, once the subject is aware that they are being observed their behaviour tends to change, since it is natural for individuals to want to make a favourable impression on others (Clark 1991; Robson 1993). There are many studies that have argued whether or not the observation and/or recording of interaction affect the behaviour of participants. Some suggest that over a period of time members become familiar with the observer and the group resumes normal behaviour. However, unless a researcher uses hidden devices or attends a meeting under the guise of a normal participant, upon which ethical arguments emerge, there are few other options available to the researcher.

## Coding communication and multiple observation techniques

Before selecting the method it is worth exploring the practical limitations associated with classification systems and multiple level observation techniques. Gorse and Emmitt (2005a,b) conducted a study of research students who were analysing group behaviour. The researchers provided comments on their experiences when quantifying communication, coding using Bales' IPA and on their use of a Simple Multiple Observation Technique. A summary of the reflections is presented in Tables 5.3, 5.4 and 5.5.

*Table 5.3* Quantifying communication acts, identifying the act, the sender and receiver

| Strengths | Limitations |
| --- | --- |
| Simple, accurate and relatively consistent statistics produced across different observers. Although quantities produced by the observers can vary, sample sizes are so large that the differences are not significant. | Recording communication acts is laborious and time-consuming. |
| | Sometimes it is difficult to know who is speaking to whom. |
| Reveals trends that are not apparent without it. | Sometime it is difficult to identify an individual recipient, so it is taken that the whole group is being addressed, rather than a particular individual. |
| Useful for identifying pairs who work together and subgroups. | |
| Provides indication of the participant's willingness to be involved. | Can be more than one intended receiver, but not broadly directed at the group. |
| Allows the researcher to look at individual and group level communication and examine who sends the data and whom it is sent to. | Impossible to capture all communication acts. |
| | If the group divides into subgroups and separate conversations take place, it is difficult or impossible to identify all communication acts. |
| The group can be split into subgroups, for example male and female and interaction examined within these categories. | |
| Allows patterns to be identified, for example who contributed throughout, who interjected periodically, who appeared to dominate the proceedings and which members were reluctant to communicate. | Frequency counts do not indicate the nature, quality, relevance or length of communication. |
| | Does not show periods of no interaction takes place. However, can be presented over time rather than cumulative to show who talks when, and when nobody talks. |
| Quantitative data can be examined over time (longitudinally), during different phases or time periods (segmental) or as one unit of data for a group, or individual (cumulatively). | Does not recognise less frequent communicators that nonetheless make a valid and possibly lengthy contribution. |
| | Fails to identify or discount verbal messages that are sent, but not received. |

*Table 5.4* Coding and categorisation of communication acts, for example Bales' IPA

| Strengths | Limitations |
|---|---|
| The group as well as the individual discussion can be broken down into different categories, which are easily analysed, in many different ways, once the data is collected. | Some interactions are difficult to classify, especially for the untrained researcher. |
| Standardisation: some methods, such as IPA, are widely recognised and can be compared to other research. | Where understanding of a communication category starts out incorrect, they will probably continue to be incorrect. |
| Can be used alongside other methods for cross comparisons and deeper understanding of behaviour, for example Belbin's team role classifications. | One act may seem to fall into two classifications and a decision needs to be made. |
| Codification correlates between different observers. | Without training to calibrate observations, the results may be inconsistent and unreliable. |
| Easy to categorise acts, after some practice. | Contributions from group members with different international origins can be difficult to classify, especially if their English is not always correct. Interpretation may be incorrect. |
| Methods, such as Bales' IPA, have survived from 1950s; this suggest they are powerful data collection methods. | Observer may have social and cultural expectations that mean that they can never be entirely objective. |
| After all the data has been collected, researchers can just concentrate on one, two or all of the categories. | No way of recording whether the message was received and understood. |
| The classification helped to identify points in the group discussion that could be investigated further using other techniques. | The method is very useful, but has little relevance unless it is combined with other methods. Other methods help to explain what happens during occurrences and trends. |
| Useful for identifying categories, for example questions and then, cross-referring to other data, for example video or transcripts, to look at whether participants openly gave information or whether it was coaxed out. | Neglects the comments of what was said. Many comments have numerous purposes and meaning, classification relies heavily on the observer's ability to judge and categorise. |
| Can be used to identify specific categories and then more detailed research on those areas and sections can be undertaken. Once patterns are identified these can be cross-referenced to locate the relevant section of audio and video data and investigated using other qualitative research methods, for example ethnomethodolgy. | Can be misleading when not fully understood. If the meaning of the speaker or intent of the message is misunderstood the classification is incorrect. |
| Can check who dominates under different categories – can be quite different to who is most talkative. | Takes time to develop natural understanding, the Bales' IPA system recommends three months training. Where relationships between observers and researchers exist, there may be bias when analysing the data. |
| Helps to understanding group dynamics. | Concentration levels can be difficult to maintain over long coding periods. Does not capture or categorise for every situation. |

*Table 5.5* Simple Multiple Level Observation Technique (SMOT)

| Strengths | Limitations |
| --- | --- |
| Simply makes use of different perspectives from different participants and researchers, using different research tools to collect data on the same topic or subject in a meaningful and manageable way. | Initially, there is some difficulty in understanding how multiple level observation techniques work. After some instruction and reading the difficulties are overcome. |
| Focused and simple triangulation. Different methods and perspectives are used to collect data from a specific episode of interaction. While using different methods and perspectives the topic and focus of the observations and records are the same. | If questions or topics are vague then participants may misinterpret them. |
| | Timing of any personal reflection by group participants is crucial; reflections vary with time. |
| Gathers multiple observations using both quantitative and qualitative tools and techniques. | Good for specific focused investigation, but inevitably misses out other issues that may be important, but not considered. |
| Provides a reliable source of data (coming from multiple points, participants, researchers, observers, and evidence collected after reviewing the video footage). | The method is flexible; data could be compiled by any combination of specialist observers, participant observers and participants' reflections of events. Clear description of the research method is needed to ensure that studies can be repeated with a good level of consistency and reliability. |
| Very useful to get a deep insight into a topic. | |
| Provides a broader understanding of what is happening within the group. | |
| In some cases views are supported and in others the views are very different, providing a more meaningful perspective. | |
| The participants' recollection of events and feeling and thoughts are important. Reluctant communicators may provide deep insight into issues, even though they appear to distance themselves from group interaction. | |

## Preliminary testing of data collection methods

Following a review of earlier studies it was evident that we needed to establish whether participants would accept the use of a method to study their interaction. We also needed to find a method that was capable of extracting reliable interaction data. A pilot study was undertaken to test the suitability of different methods for collecting interaction data. The methods tested included:

- the use of diaries to collect retrospective accounts of interaction during meetings, similar to those used by Loosemore (1996a);
- audio recordings of meetings, as used by Hugill (1999, 2001);
- the observer entering the meeting to make written observations using Bales' (1950) IPA to classify interaction, as used by Gameson (1992) and (Bellamy *et al.* 2005).

### Testing of diaries

A simple data collection sheet was designed to collect retrospective accounts of meetings, identifying who was involved in discussions, the degree of involvement each participant had during discussions, the main issues that were discussed and the nature of interaction. Contact was made with five construction participants; all of these were contractor's representatives. Initial contact was made with senior personnel within the contracting organisation. The senior managers then recommended and nominated personnel who would be able to support this study. A number of meetings took place with the representative nominated to complete the diary. These meetings were used to discuss issues such as when the sheets should be completed and the type of data that could be recorded. Once the participant understood how to complete the sheets they were asked to start recording meetings. They were asked to make records of meetings that took place between the design and the contracting team.

Feedback on the method identified a number of methodological difficulties. During initial meetings with the participants the diaries were considered to be easy to use, yet only one out of five participants completed the diary sheets. This result was surprising since regular contact was maintained with all participants, and in all of the cases support for the research project was gained from more senior members (in the participant's organisation) who encouraged the participants to complete the data sheets. Even in the case of the one participant who completed the data sheets, the quality and detail of the data reduced over time, raising concerns over the consistency and the reliability of the data.

Discussions with the participants revealed that they had neither the time nor the inclination to complete the diary sheets, thus a different strategy was needed. The researcher telephoned participants on a regular basis and asked them to recount events over the telephone, which were written into a diary. This method was far more productive than the use of diaries, although there were weaknesses. Occasionally participants were too busy to talk and had to be contacted later, or they were difficult to contact. However, in most cases regular contact was maintained. Once the participants answered the telephone, most of them spoke freely recounting events. Although the detail of events was quite rich, it soon became apparent that the accuracy of events varied. Participants would forget occurrences one week but remember them the following week. The question has to be asked: If people are remembering certain events at certain times, what are they forgetting during the telephone interview? Are they recalling the events they wish to discuss, those which remain most salient or those that had the most emotional impact? The study did not attempt to answer these questions, but recognised the limitations of this method.

### Testing of audio recording

Attempts were made to negotiate the recording of interaction in two site-based progress meetings. The management company responsible for organising the meetings was approached for their consent. Once the person organising the meeting had agreed to the research, other participants in the meeting were contacted. The majority of the participants from different organisations were apprehensive about the idea of having their conversations recorded on audiocassette. Participants wanted to know which organisation had requested the research, whom the research was for and who would have access to the recordings. The concerns were that the recordings may get into the 'wrong hands' and could then be used against an organisation. A number of participants stated that they were also concerned about the recording of inappropriate language, sensitive financial information and contractual discussions.

The process of trying to gain participant co-operation was extremely time-consuming. In one of the pilot studies it took over two weeks to make contact with the eight participants who were due to attend the meeting. In this particular case a high degree of resistance was experienced from most of the participants and two refused permission to be recorded. With the amount of resistance, and additional concern that recording may change communication behaviour, the methodology was abandoned. It is accepted that, given adequate time to develop trust between researcher and participants, it is possible to negotiate access to sensitive business settings and make audio and video recordings. At the outset of this research it was considered important to gain access to multiple project meetings and capture interaction data from those meetings. It was felt that the difficulties experienced when attempting to use audio and video would significantly reduce the ease that the researcher could gain access to multiple meetings. In this phase of the research, ethnomethodology could not be used. Ethnomethodology relies on the use of verbal records and transcripts. It is still considered important that qualitative observation is also added to quantitative data; however, the detail of such comments is considerably reduced without verbal records.

### Testing of note-taking

The final method tested was for the observer to enter the meeting and record interaction on written data sheets using the IPA technique. A main contractor was contacted and two projects were identified as suitable for the study in terms of their locality and size. The main contractor's representative was approached and the potential research project discussed. Following the main contractor's representative agreeing to the research, the remaining eight participants taking part in the management and design team meeting were contacted and their agreement sought. There was little resistance to the researcher attending and observing meetings. The researcher informed the participants that any matters deemed confidential would not be recorded or would be coded in a manner that would not allow the nature of the material to be revealed and that participants would remain anonymous. Following this initial discussion all participants were happy to allow the observer into site meetings and record observations in writing. Actors from the projects that had previously refused to let the researcher make audio recordings were again approached, the researcher asking if he could observe the meeting and make written notes: this time approvals were forthcoming. Following this experience we were relatively confident that approval to observe and record meetings using the Bales IPA method would be forthcoming for the main research programme.

### *Summary*

Difficulties were experienced with developing a methodology that allowed the collection of meaningful data. It is clear that an element of trust needs to be built between the researcher and the organisation before interactions can be observed. Difficulties were experienced when attempting to collect data using audio recordings and although we recognise that it may be possible to negotiate access to record meetings, the risk of being refused access was considered to be too high. The use of diaries also presented difficulties. Research methods that require participants to undertake activities that are not part of their normal duties may hold a low priority and may not receive sufficient attention to be usable. Loosemore (1996b) recognised this in his pilot study; the leanness of the construction industry meant that participants had little spare time and their participation in research activities had to be limited. Thus the method used must cause minimum inconvenience to those participating.

The degree of negotiation and cooperation is a determining factor in the choice of method. Participants must have the time to cooperate and agree to provide the necessary information. A danger of carrying out work that requires participants to undertake additional activities is that, when building projects have problems demanding increased commitment by the participants, the research activity may suffer a reduced priority, either reducing the detail of the data collected or failing to provide data. If the research requires all participants from different organisations to complete the diaries for the research method to be successful, considerable effort may be required to gain sufficient co-operation before the research can be undertaken.

The final method of recording interaction, using data sheets completed by the observer, was the method that met with least resistance. It should be noted that, after the first few minutes of the meeting, the parties' inquisitiveness and any defensiveness or other type of behaviour towards the researcher went unnoticed. It was felt that the researcher's presence had little effect on the nature of interaction between the other participants, although this cannot be proven.

## Testing Bales' IPA: The pilot study

Schultz (1999), drawing on the work of Schein (1987), recommends the use of clinical descriptive methods when collecting data on groups. Using such methods, researchers observe overt behaviour over a specific period, collecting information through a range of techniques. Schultz (1999) identified two specific tools for providing such feedback on group interaction: Bales' (1950) IPA and Bales *et al.*'s (1979) SYMLOG (both discussed earlier). Gameson (1992) used the IPA method to study the first meeting between clients and their potential professional advisors, classifying sequences and patterns of interaction. The aim was to determine how interaction content differed according to the profession, age and experience of the professionals. He identified the main limitations of the IPA method as an inability to measure and classify the problem being discussed. Despite these limitations Gameson concluded that the IPA was the most appropriate system to classify interaction between client and professional. Additional measures were devised by Gameson to capture the content of inter-actions. Gameson transcribed the meeting allowing grouping and quantification of words used. This provided information on the topics being raised by each professional. Using an established system, Gameson was able to contribute to the theories previously developed using the IPA method. In collaboration with others, Gameson has continued to use the IPA method to study construction interaction (Bellamy *et al.* 2005).

Bellamy *et al.* (2005) looked at the differences in communication patterns of design teams in traditional face-to-face and virtual environments. Meetings were structured and staged for the purposes of examining the differences between co-located (face-to-face) and communication in remote locations using synchronous speech and visual communication technology. Initially the discussions were captured using video footage and later transcribed and coded. Whether the participants were resistant or otherwise to the use of video equipment to record the discussion was not discussed in the research. It was however noted that the investigation was restricted to research partners, thus a relationship between the participating and the research organisation had already been formed for the purpose of research. By coding the data using Bales' IPA method, Bellamy *et al.*'s (2005) research generated graphs that summarised the interaction allowing differences between the two environments to be easily identified. Categorisation systems are particularly useful for summarising and comparing data. Tables and graphs are easily generated and analysed when interaction is in a coded form. While the use of video seems possible in staged experiments, resistance to the use of audio and video footage in *bona fide* business settings is often experienced.

Wallace (1987) found single classification methods to be too limiting and chose to draw on a number of category descriptions from established methodologies as well as creating new subject specific definitions. Bakeman and Gottman (1997) support such an approach, claiming that coding systems should be developed to suit the context and that existing methods may restrict the nature of research. When using an existing method for categorising the communication act the findings are limited to the labels and definitions of each act. The aspects of communication captured are those that are described in the method. The identification, coding of content and measurement scales used by Wallace (1987) were largely based on existing methodologies, such as Interaction Process Analysis (Bales 1950), the Bettman-Park typology (Bettman and Park 1979), Evaluative Assertion Analysis (Osgood *et al.* 1959), the General Inquirer (Stone *et al.* 1966) and 'the Gottschalk-Gleser typology' (Gottschalk 1974), together with a bespoke coding system developed for construction-based interaction. Wallace provides little explanation of how each system was adapted to develop his bespoke classification method, thus making it difficult for other researchers to use his adapted method.

As previously discussed, the IPA method provides a system for identifying and classifying interaction statements into one of twelve categories (Figure 3.2). The categories fall into one of two groups, which identify communication acts as either socio-emotional or task-based interaction. Bales' IPA can be used in sensitive business environments, which allows researchers to scientifically observe situations where accounts were previously limited to *ad hoc* quotes and hearsay. Although there are limitations with the method, it provides an excellent tool for gaining interaction data that can be compared to previous studies in construction and other research disciplines. Recent studies that have successfully used the IPA method to investigate various aspects of communication include:

- Hiltz and Turoff's (1993) study of the differenced between the agreement and the disagreement categories
- Gameson's (1992) research into construction professional and client meetings
- Jaffe *et al.* (1995), Chou (2002), Bellamy *et al.* (2005) have all used IPA to investigate issues of computer mediated communication
- Bell's (2001) investigation of patterns of communication in multidisciplinary teams.

### *Observer coding: Agreement, reliability and validity*

It is not easy to ensure that observer ratings are both valid and reliable (Clark 1991). Indeed, it is unrealistic to expect two or more researchers to break down, capture and code the continuous interaction into exactly the same categories, thus a certain amount of inter-observer variation is to be expected (Bales 1950). However, there should be a significant degree of agreement between observers, consistent with the established protocol (Clark 1991; Bakeman and Gottman 1997). The code is equivalent to a foreign language. Researchers must first learn how to use it, how to recognise and interpret the meaning of words and then translate it into the language required, or in this case the code. To ensure that codes are applied consistently, checks should be made against the original protocol and researchers using the system. When administered correctly there will be consistency in the way the data is coded, that is researchers use the same language and their interpretation of events is made without any significant differences.

It is crucial that the coding system is reliable (Berelson 1971; Mostyn 1985). Reliability is the extent that the measurement is consistent (Viney 1983); when two or more researchers code the same piece of discussion they would arrive at the same set of codes. For accurate content analysis, coding systems should be stable and reproducible (Krippendorff 1980):

- *Stability*: This is the extent to which the results of a classification are invariant over time. Each time the code is used it is applied in the same manner. Stability is achieved when the same content is repeatedly coded in the same way by the same coder.

- *Reproducibility (intercoder reliability)*: This is the extent to that multiple coders are able to classify interaction in the same manner. The interpretation of interaction and the classification is consistent across multiple coders.

With regard to reproducibility, Bakeman and Gottman (1997) make an important distinction between observer agreement and observer reliability. They argue that agreement is simply the degree that two or more researchers agree on the use of categories. Reliability is a more restrictive term and needs to be measured against a standard protocol. Observer agreement only addresses errors between observers and does not measure any deviation from a set standard. It is therefore important to have a set protocol that can be used as an index or baseline of observer reliability.

Where definitions are established and used by different researchers to categorise interaction, reliability tests are necessary to check the consistency of recording by different observers. Checks need to be carried out to ensure that coding of interaction is the same, at least within significant degrees of confidence. Variations when classifying or coding communication acts can occur in the following areas (Bales 1950: 101), although the use of systematic training will help to reduce the effect:

- *Unitising*: The identification of the discrete act; division of interaction into a number of communication acts, which are recognised and recorded separately. One utterance, statement or sentence may be coded as a single unit or may be broken down into two or more parts. It must be clear how interaction is divided and what constitutes an act suitable for coding.
- *Categorising*: The communication act or unit is identified, classified and coded. Clear definitions are important to ensure correct classification.
- *Attributing*: Identifying the designation of an originator of the communication act and the target for that communication act – who made the utterance and whom it was meant for.

### Intra- and inter-coding agreement

To some extent the level of agreement between observers coding the same piece of interaction can be verified using statistical tests. If two (or more) researchers independently and simultaneously observe and code the same piece of group interaction, tests can be undertaken to check the degree of consistency between the observations. Bales (1950) advised the use of Chi-square, rather than a correlation coefficient, to determine that the observers can score according to a specified direction. Chi-square is usually used to find a significant difference between sets; however, in this situation the statistical test is used to measure coder agreement. Bales suggested that the correlation coefficients $r$ tends to be relatively insensitive to variations in values with small densities. When coding using Bales interaction process analysis some of the categories tend to occur much less than others, they have smaller densities compared to the other categories.

When using a correlation coefficient, it is possible to find comparisons between observers which have an $r$ above 0.9, indicating high correlation, which do not come within the $\rho = 0.05$ level, indicating a significant difference, when tested by Chi-square. The Chi-square statistic is used as an indication of 'goodness of fit', being applied to a situation, which does not represent random sampling. The use of the Chi-square at the 0.50 probability level (not $\rho = 0.05$), or above, demonstrates that the system being used is common to observers and not exact (a $\rho$ value of less than 0.5 indicates an unacceptable level of difference). What is achieved is a level of agreement between coders that is not identical, but is considered suitably consistent and acceptable.

Gameson (1992) adopted the Chi-square test to ensure that categorisation was common. Initial scores were within acceptable limits ($\rho = 0.67$) and with training a reliability of $\rho = 0.89$ was achieved. Gameson's reliability tests were carried out using transcripts generated from audio recordings, that is the coding of data was retrospective. Wallace (1987) used a classification system adapted from a number of established systems, including the IPA method, but did not report any tests used to determine the ability of observers to categorise communication acts consistently. Research that fails to undertake such tests may be challenged on grounds of agreement and reliability, especially when a system does not have a proven track record.

None of the statistical tests mentioned above were developed specifically for the purpose of measuring the degree of agreement. Chi-square is not a good measure of association because its value does not provide information on the strength of the relationship between two variables (Norusis 1998). A method that is used to measure the degree of agreement between different people is the Cohen's kappa. The Cohen's kappa not only looks at the amount of ratings that are the same, but also considers the effect of agreement occurring by chance (Norusis 1998). Cohen's kappa corrects the observed per cent agreement for chance; it normalises the resulting value so that the coefficient always ranges from $-1$ to $+1$. A value of $+1$ would indicate a perfect agreement and $-1$ shows perfect disagreement. A value of 0 indicates the level of agreement that would be expected by chance. This method can be used to test whether individual communication acts are coded to a determined level of agreement, but it cannot be used to check agreement between two coders when coding continuous speech. When observers code continuous speech using IPA they identify the speech unit and classify that unit. A degree of variation would be expected for both parts. Some observers would occasionally break up speech into more communication acts than others would and code a communication act differently. Kappa can only be performed when both variables use the same category values and both values have the same number of categories; the number of observations must also be symmetrical.

If coders were given a list of 100 statements and asked to classify each statement into the appropriate IPA categories, the Kappa test can be applied. Therefore, the Kappa test can be used as a training tool ensuring that recognition of categories in a list form is reliable, and Chi-square can be used to determine that the live communication acts can be reliably categorised. Using the many examples of each of the 12 categories provided by Bales (1950), tests can be carried out on the reliability of observers to recognising and coding categories. Examples provided by Bales can be listed on a page and observers asked to identify the categories.

### Reliability and validity tests and results

Training exercises were undertaken to ensure that the data collected would be reliable and valid. The first exercise was to develop an understanding of each category and then check the degree of reliability that could be achieved when classifying communication acts against a set protocol. One hundred different statements were reproduced from Bales (1950) and these were then listed on a sheet of paper. The coder assigned an IPA category to each of the statements. Using Cohen's kappa test to examine the results against the protocol, scores in excess of 0.95 were achieved. Then an attempt was made to code the interaction of a group of postgraduate students discussing an assignment problem. This exercise was repeated a number of times with the researcher re-reading information on coding of categories between each observation. Once confident with application of the system, comparisons were made with another researcher trained in how to use the system. Independently the two researchers coded the interaction of postgraduate students engaged in a group discussion. The intercoder agreement score developed over a two-month period from $\chi^2 = 14.648$, $df = 7$, $\rho = 0.04$ (unacceptable) to $\chi^2 = 4.916$, $df = 8$, $\rho = 0.766$ (acceptable, see Appendix 2).

Following this the researcher attended three site meetings and coded the interaction. During the first meeting difficulties were experienced with the coding sheet. The coding sheet was simplified and refined for ease of use; the final coding sheet is shown in Appendix 3. With the changes being made to the coding sheet it was decided to undertake further agreement tests (Appendix 2). Two different researchers helped with this task. To enable checking of inter-coder reliability and for ease of observation, a video record of a small group interacting was used for this study. The inter-coder reliability score for the researcher and one of the assistants was much more stable and agreement levels were achieved relatively quickly. A three-month training period was required to achieve satisfactory reliability scores between all three coders, something that Bales (1950) considered to be normal. The results of the reliability tests can be seen in Appendix 2. The reliability scores were not as high as studies that used transcripts, such as those achieved by Gameson (1992), nevertheless they were within the limits considered acceptable by Bales. Following the training the observers' ability to code using the IPA method was deemed to be satisfactory.

### Obtaining a representative sample

Meetings should be representative of the project team's interaction, although consideration must also be given to the number of projects to be observed. Attempting to observe all site-based progress meetings during the construction phase on a project, typically over a period of 12–18 months, would be very time-consuming and logistically fraught with difficulties. The way around this problem is to reduce the research period by observing discrete parts of the process, that is to take a reductionism approach as used by other researchers of construction communication (e.g. Wallace 1987; Pietroforte 1992; Loosemore 1996a; Emmitt 1997; Hugill 2001; Abadi 2005). Problems of representativeness have been recognised where research is based on a small number of cases; however, there is little guidance about how to determine an appropriate number (Loosemore 1998b). It can be difficult to dictate the sample size of 'real' research projects when trying to collect data from natural and dynamic contexts (Robson 1993). For example, Simister (1995) invited 50 clients to take part in a study from which only four agreed to the proposed use of observational techniques.

Case studies are context dependent, determined by the homogeneity of the research population, the depth of the research method, time and cost limitations and experiences gained from the pilot studies (Sudman 1976; Hakin 1987; Bryman 1992; Yin 1994; Loosemore 1998b). Loosemore collected data until the patterns of communication and behaviour that emerged from analysis repeated itself. Another method is to adopt purpose sampling (Patton 1990), where samples for the research are selected based on their potential and ability for informing the research question (Azam *et al.* 1998). The degree of confidence that a sample taken from one project's meetings are representative of other meetings in that project can be checked. It is not possible to ensure that the study is representative of the general trend of interaction across all projects, but with a transparent and reliable research method further research can be undertaken to determine the extent of the trends found and the degree of application to construction and other fields.

### *Representative observations*

To achieve representative data effectively a number of observations are required to reduce the impact of unusual events. With a small sample size it is not possible to determine whether the findings are representative of the wider construction community, but the data should be representative of the projects studied. Similarly a method must be adopted that would allow others to compare their findings against this research. The data can then be used to examine whether the findings are applicable to the wider population and/or to another field.

Interaction in the same group is known to be relatively stable. There was no reason to suppose that the interaction behaviour of the management and design team on one project would vary; however, the research sought to ensure that the data collected from a series of three meetings was representative of the general nature of the meetings on a project. A pilot study was conducted to test that samples observed were representative of larger interaction sequences from the same projects.

Interaction of site-based meetings on two projects comprised the pilot study. Five meetings were observed on the first project and eight sequential meetings on the second. Tables 5.6 and 5.7 show the percentage of group communication acts for each of the IPA categories on both projects. The profiles for each of the meetings follow a similar pattern, although greater variation between meetings

Table 5.6 Pilot study 1: Observations of 5 sequential meetings

| Interaction categories | Meeting 1 No. | % | 2 No. | % | 3 No. | % | 4 No. | % | 5 No. | % | Total No. | % |
|---|---|---|---|---|---|---|---|---|---|---|---|---|
| 1: Shows solidarity | 1 | 1 | 1 | 0 | 0 | 0 | 1 | 0 | 0 | 0 | 3 | 0 |
| 2: Shows tension release | 2 | 1 | 5 | 1 | 5 | 1 | 4 | 2 | 4 | 2 | 20 | 2 |
| 3: Agrees | 21 | 10 | 32 | 8 | 19 | 6 | 9 | 4 | 8 | 4 | 89 | 7 |
| 4: Gives suggestion | 28 | 14 | 58 | 15 | 50 | 14 | 29 | 13 | 20 | 11 | 185 | 14 |
| 5: Gives opinion | 52 | 26 | 80 | 21 | 80 | 23 | 60 | 26 | 56 | 30 | 328 | 24 |
| 6: Gives information | 56 | 28 | 97 | 25 | 86 | 25 | 58 | 25 | 46 | 24 | 343 | 25 |
| 7: Asks for orientation | 12 | 6 | 40 | 10 | 30 | 9 | 16 | 7 | 12 | 6 | 110 | 8 |
| 8: Asks for opinion | 17 | 8 | 46 | 12 | 37 | 11 | 38 | 17 | 34 | 18 | 172 | 13 |
| 9: Asks for suggestion | 9 | 5 | 17 | 4 | 29 | 8 | 10 | 4 | 8 | 4 | 73 | 5 |
| 10: Disagrees | 3 | 2 | 7 | 2 | 7 | 2 | 5 | 2 | 0 | 0 | 22 | 2 |
| 11: Shows tension | 1 | 1 | 3 | 1 | 4 | 1 | 1 | 0 | 2 | 1 | 11 | 1 |
| 12: Shows antagonism | 0 | 0 | 0 | 0 | 1 | 0 | 0 | 0 | 0 | 0 | 0 | 0 |
| Total interaction observed | 202 | | 386 | | 348 | | 231 | | 190 | | 1375 | |

Table 5.7 Pilot study 2: Observation of 8 sequential meetings

| Interaction categories | Meeting | | | | | | | | | | | | | | | | Total | |
|---|---|---|---|---|---|---|---|---|---|---|---|---|---|---|---|---|---|---|
| | 1 | | 2 | | 3 | | 4 | | 5 | | 6 | | 7 | | 8 | | | |
| | No. | % | No. | % | No. | % | No. | % | No. | % | No. | % | No. | % | No. | % | No. | % |
| 1: Shows solidarity | 0 | 0 | 0 | 0 | 0 | 0 | 1 | 0 | 1 | 1 | 0 | 0 | 0 | 0 | 0 | 0 | 2 | 0 |
| 2: Shows tension release | 13 | 3 | 4 | 2 | 5 | 3 | 6 | 2 | 0 | 0 | 19 | 4 | 11 | 3 | 7 | 4 | 65 | 3 |
| 3: Agrees | 38 | 7 | 13 | 6 | 18 | 9 | 13 | 5 | 6 | 6 | 30 | 7 | 19 | 5 | 12 | 7 | 149 | 6 |
| 4: Gives suggestion | 88 | 17 | 38 | 16 | 37 | 18 | 41 | 15 | 17 | 16 | 59 | 13 | 63 | 17 | 31 | 18 | 374 | 16 |
| 5: Gives opinion | 127 | 25 | 65 | 28 | 38 | 19 | 60 | 22 | 28 | 26 | 118 | 26 | 110 | 30 | 41 | 24 | 587 | 25 |
| 6: Gives information | 162 | 31 | 69 | 30 | 55 | 27 | 94 | 34 | 33 | 31 | 148 | 33 | 109 | 30 | 45 | 27 | 715 | 31 |
| 7: Asks for orientation | 40 | 8 | 19 | 8 | 26 | 13 | 28 | 10 | 8 | 8 | 43 | 10 | 31 | 8 | 15 | 9 | 210 | 9 |
| 8: Asks for opinion | 24 | 5 | 8 | 3 | 8 | 4 | 15 | 6 | 5 | 5 | 15 | 3 | 13 | 4 | 3 | 2 | 91 | 4 |
| 9: Asks for suggestion | 16 | 3 | 10 | 4 | 11 | 5 | 13 | 5 | 6 | 6 | 13 | 3 | 9 | 2 | 7 | 4 | 85 | 4 |
| 10: Disagrees | 7 | 1 | 4 | 2 | 4 | 2 | 0 | 0 | 2 | 2 | 4 | 1 | 3 | 1 | 8 | 5 | 36 | 2 |
| 11: Shows tension | 2 | 0 | 2 | 1 | 1 | 1 | 0 | 0 | 0 | 0 | 2 | 0 | 1 | 0 | 0 | 0 | 8 | 0 |
| 12: Shows antagonism | 1 | 0 | 0 | 0 | 0 | 0 | 0 | 0 | 0 | 0 | 0 | 0 | 0 | 0 | 0 | 0 | 1 | 0 |
| Total interaction observed | 518 | | 232 | | 203 | | 275 | | 106 | | 451 | | 369 | | 169 | | 2323 | |

occurs in project one (Figures 5.1 and 5.3). The general trend with the majority of interaction being task-based (IPA categories 4, 5, 6, 7, 8 and 9) and small amounts of socio-emotional interaction (IPA categories 1, 2, 3, 10, 11 and 12) repeat itself throughout all meetings. However variations within the individual categories and between meetings do exist. To ensure that extremes of interaction are considered, but do not represent normal interaction, a minimum of three meetings should be observed for each project. This will have the effect of normalising the data, making it more comparable to the project's interaction.

A comparison was made between grouped observations of two meetings and three meetings, against the aggregated results from all of the meetings (Figure 5.2). The results demonstrate that limiting observations to only two meetings (shown by the + symbol) considerably increases the potential variation

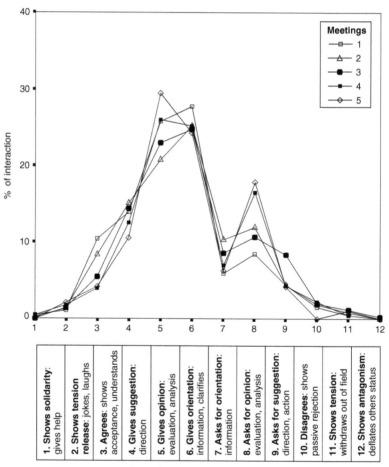

*Figure 5.1* Pilot study 1: Observations of 5 meetings

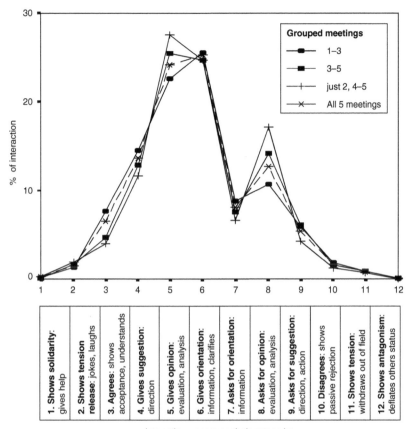

*Figure 5.2* Grouped results of 2 and 3 meetings compared with grouped results for all 5 meetings observed

from the line graph that represents all meetings. The two interaction profiles represented by three meetings (shown by the solid symbols ■) are much closer to the aggregated observations of all meetings (shown by the * symbol).

In pilot study 2 (Figure 5.3), the data represents a larger sample of the project meetings. Interaction observed in these meetings was more consistent than the first pilot study. A comparison was made between 4 sets of data, including meetings 1–3, 3–5, 5–7 and all 8 meetings. The results of both pilot studies show that extremes of behaviour are normalised by grouping project data from 3 meetings (Figure 5.4). For example, meeting 1 was the only meeting where category 12 was observed, meeting 2 had the highest level of category 11, meeting 3 had the highest occurrence of category 3 and 7 and the lowest level of category 6. However, when these were aggregated with other meeting observations the effects of the extreme values were substantially reduced.

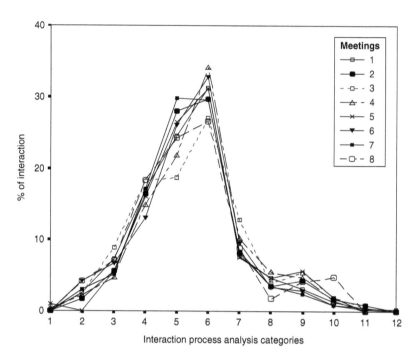

*Figure 5.3* Pilot study 2: Observations of 8 sequential meetings

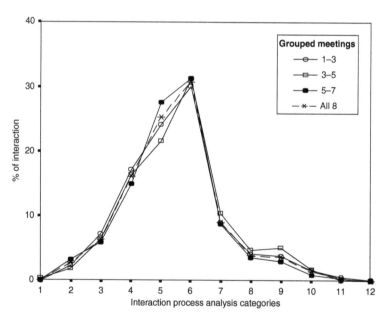

*Figure 5.4* Grouped results of 3 meetings compared with grouped results for all 8 meetings

In many cases the lines for grouped observations overlap and it is difficult to distinguish between the grouped observations for all meetings and those that only represent three meetings. The degree of commonality found shows that the patterns produced from three meetings do represent the interaction trends that exist within project group meetings.

*Summary*

Direct observations using Bales' IPA is capable of collecting data from site-based meetings. The method enables the recording of communication patterns and interaction behaviour. Variations were present when comparing meetings, although the aggregated results of three meetings on each project normalises the interaction so that it is representative of the interaction trends found in each project. We also attended meetings on different projects, first as members of the M&D team and then as researchers using Bales' IPA and were unable to find any obvious differences in the way the meetings were conducted. This tends to suggest that our presence as researchers did not adversely affect the behaviour of those present, although this cannot be determined for certain.

## Research method

A company director of a large contracting company, known to the authors, was approached and asked for his help. The contracting organisation was involved in a significant number of construction projects that were located within a reasonable geographical distance of our offices, this allowed researchers time to get to the meetings and record data during a normal working day. The contractor was a well-known organisation, involved in construction projects both regionally and nationally. After we had explained the background to the research and the proposed method of data collection, the director gave his permission for the research to be carried out and for the results to be disseminated (subject to approval from the meeting participants). This had the effect of tying the researchers into the contractor's current projects, some of which were managed using design and build, some of which were procured under a traditional contract (i.e. contract type was not controlled by the researchers). Following the meeting the director provided us with project details and contact details for the site managers. All projects observed were located in the same geographical region of the UK and were associated with commercial building projects of a varied nature. This meant that each project was different to the next and therefore the meeting participants would differ between projects.

Ten construction projects were used to gather data (summarised in Table 5.8). Eight of the projects were managed under a general contract (architect led) and the other two were managed through design and build contracts (contractor led). All projects had a minimum contract value of 3 million pounds and a

*Table 5.8* Description of the construction projects

| Project number | Project description | Approximate project value (£ millions) | Procurement method |
|---|---|---|---|
| 1 | Two-storey new build call-centre with parking facilities. | 7.5 | Design and build |
| 2 | Conversion of existing inner-city building into luxury flats and restaurant. | 4.5 | General contract |
| 3 | Eight-storey block of luxury flats with parking facilities. | 5.0 | General contract |
| 4 | Inner-city four-storey building converted to luxury flats, part new build and part refurbishment. | 4.5 | General contract |
| 5 | Two-storey medical centre, shops and public amenities. | 4.5 | Design and build |
| 6 | New school building with parking facilities. | 7.0 | General contract |
| 7 | Five-storey building providing accommodation for luxury flats and restaurant part new build, part refurbishment. | 4.8 | General contract |
| 8 | Refurbishment of four-storey building to accommodate flats and restaurant. | 3.5 | General contract |
| 9 | Refurbishment and new build to hospital wing. | 3.5 | General contract |
| 10 | Refurbishment of five-storey building to accommodate a recording studio and media centre. | 5.2 | General contract |

maximum contract value of 8 million pounds sterling. The total contract value of the projects studied was approximately £50 million. All observations were undertaken during the construction phase when professionals interacted in site-based progress meetings. The site-based progress meetings were all held in the contractor's site accommodation offices. The interaction of each participant at the meeting was observed and coded using Bales IPA. By gathering data from all contributors it was possible to aggregate the interaction results to provide a profile of group interaction. Because performance information was collected on the contractor's representative, specific attention was also given to the observations of interpersonal interaction between the contractor's representatives and other members of the management and design team. Interpersonal profiles for the contractor's representative were also produced.

## Use of the IPA method

Data were gathered from each of the construction projects by attending, observing and recording interaction from three to four sequential meetings (discussed in more detail below). In total, data were gathered from 36 site meetings, involving 96 different participants (listed in Table 6.1). Before observing the first progress meeting on each project the researcher contacted all the key personnel scheduled to attend and briefly outlined the nature of the research. All of the subjects were shown the data collection sheet (Appendix 3). Reassured by the type of data that was to be collected, all of the participants gave their consent for the meetings to be observed and the data recorded on the data collection sheets. During the first meeting of every project the researcher was briefly introduced to the meeting participants. At the start of the meeting some of those present asked to have a look at the data collection sheet, which was shown to them. Following this, those attending the meeting paid little attention to the presence of the observer.

At the end of the contract period, and following the observation of all case studies, data on the success of the project was collected from the managing directors of the main contracting organisation. This provided enough information to determine whether the project had been completed to budget and to programme. The criteria for evaluating project success were:

- Project completion within or in excess of the scheduled duration;
- Project completion within or in excess of the estimated budget;
- Whether or not conflict occurred between the contractor and the architect;
- Whether or not conflict occurred between the contractor and the client.

Half of the projects were completed on time and within budget. A further two were completed within budget, but experienced time overruns. Three of the ten projects failed to complete on time or within budget. Conflict with the client was witnessed on three of the ten projects and only one project revealed conflict with the architect (Case study 6). The five projects that were completed on time and within budget did not demonstrate any conflict with the client or architect, and could be considered to be successful projects. Two of the projects that failed to deliver on time or programme also experienced conflict with the client, while the third experienced conflict both with the client and the architect and is thus classified as the least successful project (Table 5.9).

Following the completion of the observation period we asked the company directors to rate the effectiveness of the contractor's representatives. Their evaluation was based on the site manager's previous ability to contribute to successful projects, that is those that were profitable, completed on time and did not result in disputes. This allowed a comparison to be made between the perceived effectiveness of individual site managers and their performance as measured during the research period.

*Table 5.9* Project performance

| Project number | Completed within budget | Completed within scheduled programme | Experienced conflict with the client | Experienced conflict with the architect | Number of positive outcomes |
|---|---|---|---|---|---|
| 1 | Yes | Yes | No | No | 4 |
| 2 | Yes | No | No | No | 3 |
| 3 | Yes | Yes | No | No | 4 |
| 4 | No | No | Yes | No | 1 |
| 5 | Yes | Yes | No | No | 4 |
| 6 | No | No | Yes | Yes | 0 |
| 7 | No | No | Yes | No | 1 |
| 8 | Yes | No | No | No | 3 |
| 9 | Yes | Yes | No | No | 4 |
| 10 | Yes | Yes | No | No | 4 |

## Data processing and analysis

The interaction observed during meetings was manually recorded using a tabulated data sheet (Appendix 3). Information on the length and nature of communication acts is provided in Appendix 1. Data collected at each meeting included the identification of the:

• Participant who was initiating communication;
• Participant at whom the communication is predominantly aimed; and the
• Communication act using the 12 socio-emotional and task-based categories provided by Bales' IPA method.

Reflective qualitative notes were also made following the meeting to aid interpretation of the IPA results. These were based on interaction behaviour that would not be identified by the IPA system and were limited to overt interaction behaviour between subjects. The reflective data adds a contextual dimension to the IPA data helping explain the nature and use of communication acts. For example, although the IPA system would identify where conflict, disagreements and disputes occurred it would not identify the nature of that conflict within the group discussion, nor would it identify the intensity and how it was used. By making notes after the meeting the observations made during the meeting and the IPA coding were not compromised.

The IPA quantitative data were manually entered into a computer-based spreadsheet (SPSS v. 10) ready for statistical analysis. The data relating to the project's success and the effectiveness of the contractor's representatives were also entered into the spreadsheet.

Investigation of the first research objective was undertaken by analysis of the quantitative data on a case-by-case basis in the form of graphical profiles and tabulated results. Then the aggregated results for all observations were used to provide a single profile for the management and design team meeting. This profile could then be compared to previous studies of groups based on the IPA method. Testing the validity of the first hypothesis is presented in two formats:

1. Analysis of quantitative data presented in the form of graphical profiles and tabulated results allows comparison between the project's success criteria;
2. Statistical analysis is then undertaken, using the Chi-square statistic, to determine if differences found are significant.

Testing the validity of the second hypothesis is presented in two formats:

1. Analysis of the quantitative data in the form of graphical profiles and tabulated results allows comparison between the contractor's representatives rated effective and those rated less effective;
2. Statistical analysis, using the Chi-square statistic, was used to determine if different interaction profiles based on the effectiveness of the contractor's representatives were significantly different.

## Coding and analysing Bales' IPA

The categories that are used to code communication are identified in Figure 3 (further information is also provided in Appendix 1), which provides a very brief outline of the 12 categories prescribed in the original protocol. A more detailed description of each category can be found in Bales (1950: 177–195). Appendix 1 also provides guidance on what constitutes a full speech act.

### Method of categorising communication

Each communication act was coded in accordance with Bales' IPA. The smallest part of the interaction process coded and analysed was a sentence or a statement of meaning. Analysis of individual words would be impractical when observing and recording interaction in real time, and would be of little use when analysing the results. The person speaking and the direction of interaction, the person being addressed, are also recorded.

### Analysing the data

It is conventional for researchers using the Bales methodology to pay particular attention to the trends found in the descriptive statistics; using line graphs (profiles) and tables, indeed many including Bales choose not to support all of their analyses with inferential statistics. Most recent IPA studies use a combination of descriptive and inferential analysis (Cline 1994; Socha and Socha 1994; De Grada *et al.* 1999). Despite Halpin's (1990) use of various statistical methods he suggests that problems in comparing IPA group profiles on statistical analysis alone centres around a few related issues:

- The differences between 2 sets of data across all 12 IPA categories are typically small. Many statistical tests are inappropriate for such small differences.
- Some of the categories tend to be much larger than others, routinely accounting for 30–40 per cent of the total IPA, while others that are equally significant are seldom scored. In most observations the task-based categories account for the majority of the interaction, and although the socio-emotional categories are small, they often have the greatest effect on the group.
- Only some IPA categories are considered to be relevant to a particular trait or association.

The results presented here first discuss the descriptive trends and relationships with the aid of qualitative notes, line graphs and tables before analysing differences using inferential statistics. The statistics and tables are produced with the aid of a statistical processing package, namely SPSS version 10. The Chi-square test will be used to determine if differences are significant. The IPA system relies on the classification of communication from a finite number of nominal communication categories (12). The Chi-square test is used to see if there is evidence that suggests that two categorical variables are associated with each other, determining if they are related in population. The Chi-square test does not measure the strength of the relationship, it measures whether the relationship occurred due to chance.

Once it is determined that differences between data sets are significant the adjusted residuals can be used to examine where the greatest differences occur. Residuals are the difference between the observed value and the expected value. The expected value is the number of cases one would find if there were no relationship between the variables. Positive residuals indicate that there are more cases in the cell than there would be if the row and column variables were independent. The residuals are then divided by the estimate of its standard error and expressed in standard deviation units above or below the mean. This is known as the adjusted residual.

### Tabulated results

All of the results are presented in tables and graphically using line graphs. Line graphs are traditionally used to present IPA results. The use of line graphs allows multiple sets of data to be laid over the top of each other allowing easy comparison. Because there are 12 interaction categories, comparing multiple sets of data using histograms can result in the graphical representation becoming cluttered and difficult to read. The tables used (e.g. Table 6.1) identify the number of communication acts observed in each category (No.), the percentage of interaction, calculated vertically, and the adjusted residual (Adj.).

## Multiple comparison of the 12 IPA categories

The group interaction data was analysed from a number of different perspectives: whether projects were completed within the scheduled duration, within budget, whether conflict was experienced between the contractor and client, and contractor and architect. Each analysis will consider the 12 IPA categories separately and when grouped into socio-emotional and task-based headings. When using multiple comparisons, there is a need to know that all probability statements are simultaneously correct; therefore, confidence statements should be adjusted to reduce the potential for error. Undertaking multiple levels of analysis on one data set can result in Type 1 error (rejecting the null hypothesis when it is true). The chances of finding significant results are inflated as the number of tests increase. However, to reduce the potential problems associated with multiple comparison $p = < 0.05$ should be treated with care, with $p = < 0.01$ being a more reliable predicator (also $p = < 0.001$ is still highly significant). Lederman (1984a,b) suggests the use of the Bonferroni method (dividing the confidence level by the number of multiple comparisons) of constructing simultaneous confidence intervals. Thus, when analysing the 12 individual categories a confidence level of $p = 0.05$ should be divided by 12 $(0.05/12 = 0.004)$. Undertaking so many comparisons, with large amounts of data, may result in findings of such small magnitude. Following the analyses of results, attention needs to be paid to the descriptive results and the degree and direction of differences identified in the data.

# 6 Interaction data from construction meetings

The interaction data collected from the management and design (M&D) team meetings are presented here. Line graphs and result tables provide the descriptive statistics for each individual meeting. The IPA data are shown below, with all percentages rounded up for ease of reading. Rounding up does produce a slight error (not all percentages add up to 100 per cent); however, all data counts are included so that it is possible to calculate more precise figures if required. Comments are made on the nature of interaction immediately prior to the meetings and changes in personnel attending the meeting are also noted. An initial analysis of the trends and differences are reported after the results, these are supported with qualitative observations to help contextualise the quantitative data. The inferential statistics and analysis regarding successful and unsuccessful project outcomes are presented in the following chapters. Success of the project team was determined by the completion of the project on time and within budget.

The contractors' directors were also asked whether any conflict had emerged between the contracting parties. Conflict in this instance is referring to formal or legal action taken by the parties as a result of failing to resolve matters on site, it does not refer to conflict that may manifest within meetings as a part of problem resolution. Conflict, when used as a measure of unsuccessful performance, is only considered when disputes continue after the building project has been completed, are formal involving the exchange of written documentation or when third parties are engaged to assist in the resolution of such matters (e.g. legal consultants, lawyers, mediators, adjudicators, arbitrators etc.).

Table 6.1 provides a summary of the professionals and the number of people attending each meeting. The highest number of people attending a meeting was 11 and the lowest 3. Rounded up, 7 were the average number of people that attended the meetings and 9 were the mode. Although there are often many more professionals involved in a construction project, it is notable that numbers attending the meetings are consistent and tend to be small. There were a core group of professionals present at most meetings, others attended to suit project requirements.

## Results: Case study 1

The first case study was a new build call centre with parking facilities; the approximate contract value was £7.5 million. The project was managed under a

*Table 6.1* Number of participants attending each meeting

| Project no. | 1. | | | | 2. | | | 3. | | | 4. | | | 5. | | | | 6. | | | | 7. | | | | 8. | | | | 9. | | | 10. | | | |
| --- |---|---|---|---|---|---|---|---|---|---|---|---|---|---|---|---|---|---|---|---|---|---|---|---|---|---|---|---|---|---|---|---|---|---|---|---|
| Meeting ID / Project participants | 1 | 2 | 3 | 4 | 1 | 2 | 3 | 1 | 2 | 3 | 1 | 2 | 3 | 1 | 2 | 3 | 4 | 1 | 2 | 3 | 4 | 1 | 2 | 3 | 4 | 1 | 2 | 3 | 4 | 1 | 2 | 3 | 1 | 2 | 3 | 4 |
| Architects | 2 | 2 | 2 | 2 | 1 | 1 | 1 | 1 | 1 | 2 | 2 | 3 | 1 | 1 | 1 | 1 | 1 | 2 | 1 | 1 | 1 | 1 | 1 | 1 | 1 | 2 | 1 | 1 | 2 | 2 | 2 | 1 | 1 | 1 | 1 | 1 |
| Contractor's site managers | 1 | 1 | 1 | 1 | 1 | 1 | 1 | 1 | 1 | 1 | 1 | 1 | 1 | 2 | 2 | 2 | 2 | 1 | 1 | 1 | 1 | 1 | 1 | 1 | 1 | 1 | 1 | 2 | 2 | 2 | 1 | 1 | 2 | 1 | 1 | 1 |
| Contractor's quantity surveyors | 1 | 1 | | | 1 | 1 | 2 | 2 | 2 | | 1 | 1 | 2 | 1 | 1 | 1 | | 1 | 1 | 1 | 1 | 1 | 1 | 1 | 1 | 1 | 1 | 1 | 1 | 1 | 1 | 1 | 1 | 1 | 1 | |
| Structural engineers | 1 | 1 | | | 2 | 2 | 2 | 2 | 2 | 1 | 1 | 1 | 1 | | | 1 | 1 | 1 | 1 | 1 | 1 | 1 | 1 | 1 | 2 | 2 | 1 | 1 | 1 | 1 | 2 | 1 | 1 | 2 | 1 | 1 |
| Contracts' managers | | | | | 1 | 1 | 1 | 1 | 1 | 1 | | | | | | | | 1 | 1 | 1 | 1 | 1 | 1 | 1 | 1 | 1 | 1 | 1 | 1 | 1 | 2 | 1 | 1 | 2 | 1 | 1 |
| Client's quantity surveyors | | | | | 1 | 1 | 1 | 1 | 1 | 1 | 1 | 1 | 1 | | | | | 1 | 1 | 2 | 1 | 1 | 1 | 1 | 1 | | | | | 1 | 1 | 1 | | | 1 | 1 |
| Mech and Elec consultants | | | | | 1 | 1 | 1 | 1 | 1 | 1 | | | 1 | 2 | 2 | 2 | 1 | 1 | 2 | 2 | 2 | 1 | 1 | 1 | 1 | 2 | 1 | 1 | 1 | | | | 1 | 1 | 1 | 1 |
| Mech and Elec sub-contractors | | | | | | | | | | | | 2 | 1 | 2 | 2 | 3 | 1 | | | | | 1 | 1 | 1 | 1 | | | | | | | | | | | |
| Client's representatives | | | | | 1 | 1 | 1 | 1 | 1 | 1 | 1 | 1 | | | | | | 1 | 2 | 2 | 2 | 1 | 1 | 1 | 1 | 1 | 2 | 2 | 2 | 4 | 3 | 1 | | | | |
| Client's project managers | | | | | 1 | 1 | | 1 | 1 | 1 | | | | | | | | | | | | | | | | | | | | 1 | 1 | 1 | 1 | 1 | 1 | 1 |
| Clerk of works | | | | | | | | | | | | | | 1 | 1 | 1 | 1 | 1 | | | | | | | 1 | | | | | | | | | | | |
| D&B co-ordinators | | | | | | | | | | | 1 | 1 | | 1 | | | | 1 | | | | | | | | | | | | 1 | | | | | | |
| Planners | | | | | | | | | | | | | | | | | | | | | | | 1 | | | | | | | | | | | | | |
| Site engineers | | | | | | | | | | | | | | | | 1 | | | | | | | | | | | | | | | | | | | | |
| Total participants | 5 | 5 | 3 | 3 | 9 | 8 | 9 | 9 | 10 | 10 | 6 | 10 | 7 | 9 | 8 | 9 | 8 | 10 | 8 | 9 | 5 | 7 | 9 | 8 | 9 | 5 | 9 | 9 | 11 | 10 | 9 | 7 | 10 | 10 | 9 | 9 |

design and build contract. It was completed within budget, on time and without conflict.

### Observation of meetings 1.1–1.4

Very little social interaction was evident between the professionals either before or after the meetings. The majority of the interaction that occurred before the meeting was contained within subgroups; for example, the senior architect would talk to the other architect, while the contractor's quantity surveyor would talk to the contractor's project manager. Conversations within the subgroups were neither loud nor whispered, these 'polite' discussions appeared to help maintain interpersonal status. The structural engineer did not make much effort to interact with the other professionals before the meeting, nor did others attempt to make conversation with him. While waiting for the quorum to gather, actors sat in silence and organised their notes and documents. The meeting was conducted in a formal and structured manner. The lack of social and informal interaction across the professional groups prior to the meeting seemed to emphasise the formality and importance of the meetings.

### Analysis

The first three meetings followed a similar profile (Figure 6.1 and Table 6.2). Most of the interaction in this and the other cases are associated with the emotional, neutral (task-based) categories. The low levels of socio-emotional interaction are consistent with earlier observations of adult groups.

Meeting 1.4 has the largest variation in the task elements, when compared with the other meetings, but this appears to be related to the fact that it was the last formally scheduled meeting. During meeting 1.4 it was stated that many of the M&D problems on the meeting agenda had been successfully resolved and even though the contract had just left to run over two months, those present were asked by the contractor's project manager whether there was a need to continue with the formal meetings. It was proposed that if any issues arose at a future date, a meeting could be scheduled and it was agreed that meeting 1.4 would be the last formally scheduled M&D team meeting.

Analysis of changes in behaviour between the meetings reveals a number of apparent trends. Ignoring the last meeting, the largest gradual variation occurred in categories three and four. Giving suggestions, direction and generally implying autonomy increased at each subsequent meeting, and there was also an increase in the level of agreement. Generally, there were more suggestions and agreements in later meetings, which is matched with a decline in evaluation and explanation. A reduction in the information requested and disagreements is also evident. The amount of tension release and tension shown is stable in most meetings, although highest in the first meeting. There was little variation in the use of showing solidarity, raising other's status, giving help and asking for opinion, evaluation

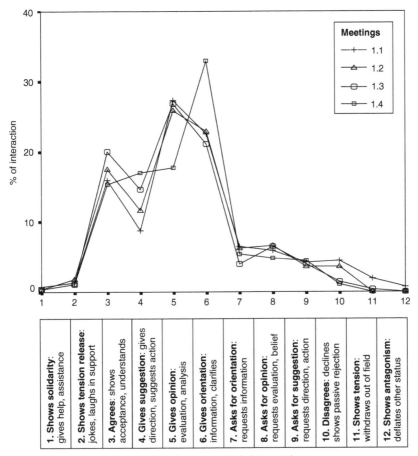

*Figure 6.1* Group interaction profiles of Project 1, meetings 1.1–1.4

or analysis. However, very few observations of showing solidarity were collected in any of the cases.

The IPA categories can also be examined in terms of task-based and socio-emotional interaction. In all four meetings the percentage of positive socio-emotional interaction (categories 1–3) was higher than negative socio-emotional interaction (categories 10–12). The total positive emotional interaction is five times higher than the negative emotional interaction. This case study has the highest overall occurrence of category three, accounting for 17 per cent of the group's total interaction.

The amount of socio-emotional interaction (categories 1–3 and 10–12) is low compared with task-based interaction. The task-based interaction in case study one accounts for 78 per cent of the interaction. Another noticeable trend is that the

Table 6.2 Results of group interaction for meetings 1.1–1.4

| Interaction categories | Meeting 1.1 No. | % | Adj. | 1.2 No. | % | Adj. | 1.3 No. | % | Adj. | 1.4 No. | % | Adj. | Total No. | % |
|---|---|---|---|---|---|---|---|---|---|---|---|---|---|---|
| 1: Shows solidarity | 4 | 1 | 0.9 | 1 | 0 | −0.3 | 1 | 0 | −0.3 | 1 | 0 | −0.4 | 7 | 1 |
| 2: Shows tension release | 8 | 1 | 0.1 | 5 | 2 | 0.8 | 3 | 1 | −0.4 | 3 | 1 | −0.5 | 19 | 1 |
| 3: Agrees | 94 | 16 | −0.8 | 48 | 18 | 0.3 | 56 | 20 | 1.6 | 45 | 15 | −0.8 | 243 | 17 |
| Total positive socio-emotional | | | | | | | | | | | | | | 19 |
| 4: Gives suggestion | 52 | 9 | −3.3 | 32 | 12 | −0.3 | 41 | 15 | 1.4 | 50 | 17 | 2.9 | 175 | 12 |
| 5: Gives opinion | 162 | 27 | 1.7 | 71 | 26 | 0.4 | 75 | 27 | 0.8 | 52 | 18 | −3.2 | 360 | 25 |
| 6: Gives information | 133 | 23 | −1.4 | 63 | 23 | −0.7 | 59 | 21 | −1.5 | 97 | 33 | 3.8 | 352 | 24 |
| Total giving task-based | | | | | | | | | | | | | | 62 |
| 7: Asks for information | 38 | 6 | 1 | 17 | 6 | 0.4 | 11 | 4 | −1.4 | 16 | 5 | −0.2 | 82 | 6 |
| 8: Asks for opinion | 35 | 6 | 0 | 18 | 7 | 0.5 | 18 | 6 | 0.4 | 14 | 5 | −0.9 | 85 | 6 |
| 9: Asks for suggestion | 25 | 4 | 0.2 | 10 | 4 | −0.4 | 11 | 4 | −0.2 | 13 | 4 | 0.3 | 59 | 4 |
| Total requesting task-based | | | | | | | | | | | | | | 16 |
| 10: Disagrees | 26 | 4 | 2.6 | 10 | 4 | 0.7 | 4 | 1 | −1.7 | 3 | 1 | −2.2 | 43 | 3 |
| 11: Shows tension | 11 | 2 | 3.6 | 0 | 0 | −1.7 | 1 | 0.4 | −1.0 | 0 | 0 | −1.8 | 12 | 1 |
| 12: Shows antagonism | 4 | 1 | 2.4 | 0 | 0 | −1.0 | 0 | 0 | −1.0 | 0 | 0 | −1.0 | 4 | 0 |
| Total negative socio-emotional | | | | | | | | | | | | | | 4 |
| Total no. | 592 | | | 275 | | | 280 | | | 294 | | | 1441 | |

Note

Adj. = Adjusted residuals – standardised difference between observed and expected.

giving of information, opinions and suggestions were considerably higher than requesting information, opinions and suggestions. The increase in positive and the reduction in negative socio-emotional interaction corresponded to a reduction in the number of issues and problems raised.

Early observations of this project witnessed a number of critical discussions, with problems listed on the agenda discussed in detail. The contractor's representatives would often ask for extra information and explanation rather than just relying on the initial response given by a specialist. The contractor's representatives would also be prompted to show when they did not agree with a suggestion. During early meetings the levels of suggestions given were high; it was only when the number of problems being raised reduced that the number of directions given also reduced.

The contractor's project manager was particularly dominant during discussions. Although a period of open discussion did take place on most issues, the project manager was firm in his insistence on bringing matters to a close and coming up with a definite course of action. If suggestions were not forthcoming he would prompt others to provide proposals and make suggestions. When proposals were not given, he would draw together information from the discussion, rephrase what he thought others were suggesting and put the idea to the group. If others agreed he would ask them to provide a firm date for completion of the suggested activities. When professionals appeared to be vague on dates (e.g. 'by the middle of next month') the project manager was quick to find a corresponding date in his diary, and confirm the date with those suggesting the time-scale. This often encouraged others to give slightly more thought to the date of the activity; however, at the end of discussions a firm date was always set, agreed and recorded in the minutes. The act of setting and recording exact dates always made those responsible for the action appear slightly anxious. Providing a set date encouraged them to give more thought to activities and the timing of events.

### Summary

All of the performance outcomes on this project were successful. The company's director gave the contractor's project manager the highest rating. The project manager chaired the meeting. A simple agenda was produced for each meeting and minutes from the previous meeting were tabled. The minutes concisely stated the actors present, apologies, what had been discussed and what action was decided and by which date. If the action had not been achieved by the next meeting, this was recorded and a new date for the action was set. At each meeting the action not completed was discussed. Those who had failed to achieve their action point were asked to provide a new date for the action to be completed. Professionals were never allowed to be vague on the date for completion. By checking the calendar and taking time to record the new date and issues discussed, it was always made clear that the date and any programme slippage were being recorded and monitored. When an actor stated that work would be completed by the end of the week, the project manager, who was chairing the meeting, opened his diary, checked the date that corresponded with the end of the week and asked if this was the date they expected to have the work completed. The tone of the conversation was factual and neutral in emotion, no positive or negative emotion was expressed. Emphasis was on business, ensuring that matters were properly considered and formally recorded. Rarely was there any detailed discussion about why the action had not taken place. The project manager never forced a date on other parties, he merely made it clear that the date had passed and that the new date was either critical or in danger of becoming critical. If the delay had affected other tasks, this was also mentioned and recorded, but again not discussed in detail. The focus on formality seemed to emphasise potential consequences, without needing to discuss them, and resulted in action. Interaction was neutral in emotion; however, the atmosphere was slightly tense and the discussions showed that matters were important.

At all stages in the discussion, the project manager let actors make their response to problems and recorded their suggestions. If there appeared to be faults in the design or poor performance of a product was suspected, the project manager would bring the matter to the attention of the relevant representative. At the meeting the project manager would not tell the representative what action would need to be taken, but would ask how they intended to resolve the matter. The project manager would outline the issue, state his concerns and would ask the relevant party for their intended action. Considerable emphasis was placed on how the party at fault intended to resolve the matter and when it would be resolved. If the project manager was concerned about the suggestion made to rectify the situation, he would probe for more information. Discussions were not prolonged and once action had been recorded matters would be moved on to the next item on the agenda.

When disagreements emerged these were often followed by questions, information seeking and probing statements that were specifically tied to the item of disagreement. Once the problem had been exposed suggestions for action were requested from those with relevant knowledge. Disagreements were swiftly resolved with suggestions of action. Suggestions were reiterated by the chair and confirmed for the purpose of the minutes, the chair would then ask for agreement and the action was formally recorded.

## Results: Case study 2

This project was a conversion of an inner-city building into luxury flats and a restaurant. It was managed under a general contract and completed within budget, but not within the scheduled programme. There was no evidence of conflict on this project.

### *Observation of meetings 2.1–2.3*

There was a considerable amount of social interaction before the start of the meeting; most of this was initiated by the client's project manager and focused on less senior representatives. The discussions tended to focus on sports events, such as rugby and football, or television and news (discussing issues of current interest). The members participating in these conversations tended to be limited to the client's representatives, including the structural engineering, mechanical and electrical engineering consultants and the client's project manager. The contractor's representatives often entered the meeting just before it was due to start, which meant that there was little time for informal discussions involving the contractor's representatives prior to the meeting. The architect tended to arrive just after the contractor's representatives, and was also not involved in discussions prior to the meeting. The meeting commenced as soon as the architect arrived.

## *Analysis*

Interaction associated with the more extreme socio-emotional categories, such as shows solidarity, raises others status, gives help and shows antagonism, deflates others status, were not observed during any of the three meetings (Table 6.3). Showing tension release, laughing and joking was only observed on one occasion, the only category remaining that provided positive emotional communication was agreement.

Case study two has greater variation between group meetings than that was found in any of the other cases. The most consistent category throughout the series of meetings is asking for opinion, analysis or evaluation. The amount of variation found in this set of meetings was particularly unusual when compared to the other cases. A considerable variation occurred in the personnel attending each of the meetings and changes in interaction patterns may be linked to changes in actors.

Meeting 2.3 was the last formally scheduled M&D meeting of this construction project. The profile for meeting 2.3 is similar to that of 1.4 (Figures 6.1 and 6.2); both of these profiles represent the last formally scheduled M&D team meetings. The only obvious difference between the two cases is that in profile 2.3 the amount of disagreements is considerably higher than that observed in meeting 1.4. This project did not complete on time and there were discussions relating to this fact.

Summarising all of the meetings, the amount of positive socio-emotional communication is almost the same as the negative socio-emotional communication. The positive exceeded the negative socio-emotional communication in only one of the meetings (meeting 2.2). Although the profiles have the greatest variation between meetings, there was still a typically high level of task-based interaction compared with socio-emotional interaction.

A degree of tension was attributable to the relationship between the client's project manager and the architect. This seemed unusual as both consultants represented the client. The critical debate between the two parties did mean that issues raised were considered from two differing perspectives. When the conflict became prolonged other group members would attempt to disperse the tension and resolve differences.

Table 6.3 Results of group interaction for meetings 2.1–2.3

| Interaction categories | Meeting 2.1 | | | 2.2 | | | 2.3 | | | Total | |
|---|---|---|---|---|---|---|---|---|---|---|---|
| | No. | % | Adj. | No. | % | Adj. | No. | % | Adj. | No. | % |
| 1: Shows solidarity | 0 | 0 | 0 | 0 | 0 | 0 | 0 | 0 | 0 | 0 | 0 |
| 2: Shows tension release | 1 | 0 | 1.3 | 0 | 0 | −1.0 | 0 | 0 | 0 | 1 | 0 |
| 3: Agrees | 11 | 4 | −3.4 | 39 | 10 | 2.7 | 11 | 10 | 0.9 | 61 | 8 |
| Total positive socio-emotional | | | | | | | | | | | 8 |
| 4: Gives suggestion | 25 | 8 | 0.7 | 22 | 6 | −1.7 | 12 | 11 | 1.4 | 59 | 7 |
| 5: Gives opinion | 145 | 48 | 4.5 | 140 | 37 | −0.5 | 16 | 14 | −5.6 | 301 | 38 |
| 6: Gives information | 47 | 15 | −4.3 | 106 | 28 | 2.7 | 35 | 31 | 2.0 | 188 | 24 |
| Total giving task-based | | | | | | | | | | | 69 |
| 7: Asks for information | 2 | 1 | −3.3 | 21 | 6 | 3.4 | 3 | 3 | −0.4 | 26 | 3 |
| 8: Asks for opinion | 22 | 7 | 1 | 21 | 6 | −0.7 | 6 | 5 | −0.4 | 49 | 6 |
| 9: Asks for suggestion | 21 | 7 | 0 | 24 | 6 | −0.6 | 10 | 9 | 0.9 | 55 | 7 |
| Total requesting task-based | | | | | | | | | | | 16 |
| 10: Disagrees | 23 | 8 | 1.4 | 8 | 2 | −4.4 | 17 | 15 | 4.4 | 48 | 6 |
| 11: Shows tension | 9 | 3 | 2.6 | 0 | 0 | −3.3 | 3 | 3 | 1.1 | 12 | 2 |
| 12: Shows antagonism | 0 | 0 | 0 | 0 | 0 | 0 | 0 | 0 | 0 | 0 | 0 |
| Total negative socio-emotional | | | | | | | | | | | 8 |
| Total no. | 306 | | | 381 | | | 113 | | | 800 | |

Note
Adj. = Adjusted residuals − standardised difference between observed and expected.

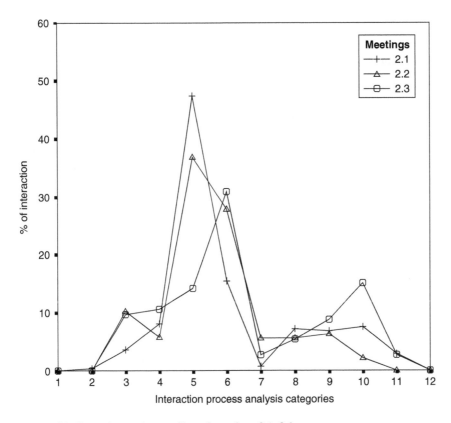

*Figure 6.2* Group interaction profiles of meetings 2.1–2.3

## Summary

The tension that manifest between the architect and the client's representative did not distract from issues discussed at the meeting. At each meeting a simple agenda was produced and additional information and representatives' reports were tabled. The meeting was conducted in a relatively relaxed environment although the chair did ensure that each matter was dealt with. Potential delays were discussed, but were not systematically recorded in the minutes of the meetings. The date when the next activity was expected was discussed and activities scheduled on bar charts noted, but no specific records of slippage were made. The client's project manager chaired the meeting. The meetings were relatively informal and unstructured; however, if matters were losing their focus, the chair was quick to end discussions and bring the focus back on the item being discussed.

## Results: Case study 3

This project comprised an eight-storey high block of luxury flats with parking facilities, managed under a general contract. The work was completed on time and within the scheduled programme. There was no evidence of formal conflict or legal disputes once the contract was completed.

### Observations of meetings 3.3–3.4

Informal discussions would take place between the construction professionals prior to the start of the meeting. The client's project manager initiated most of the discussions. The architect and the contractor's representatives arrived just before the start of the meeting, so they did not take part in these conversations. Most of the representatives who took part in the conversations were less senior members of the group. The number of people participating in the meeting was high compared with the other projects (Table 6.1).

### Analysis

The three line graphs follow a similar profile (Figure 6.3). In all of the ten case studies the line graphs within each case study typically follow similar profiles, apart from case study two. In this case study, the largest variation occurs in category 11 of meeting 3.1 (Table 6.4). Agreeing also has a high adjusted residual; this indicates considerable variation in this interaction category between meetings. However, the profiles are still typical of all other profiles with the task-based communication accounting for the majority of interaction.

The qualitative observations during the meeting noted a conflict between two of the participants. This was supported by relatively small increases in the quantitative levels of negative emotional interaction. This helps to show that slight deviations in the total quantitative measures can represent quite notable differences in the real group environment.

From a qualitative perspective the level of tension within the meetings was higher and more intense than that found in the other cases. Much of this was attributable to conflict and disagreement between the client's project manager and the architect. Even though the qualitative observation of interpersonal behaviour found the amount of negative emotional exchange to be high, it only resulted in a few instances of antagonism (category 12). The interactions between the senior architect and the client's project manager were occasionally emotionally charged, with few attempts being made to understand the other person's reasoning. The tension emerged from the advice and suggestions provided by one party, which would not be accepted by the other. The architect did not seem to like the client's project manager making suggestions that impacted on the architect's role and duties. The architect argued and repeatedly put forward the view that there were certain ways of dealing with other parties, such as statutory authorities, claiming that this is 'the way things are done'. The client's project manager often challenged this view, being the main protagonist of many critical discussions. The initial points made by the project manager were not presented as a criticism of the architect nor delivered with negative emotion; however, the architect became irritated when the project manager raised such issues and the project manager

Table 6.4 Results of group interaction for meetings 3.1–3.3

| Interaction categories | Meeting | | | 3.2 | | | 3.3 | | | Total | |
|---|---|---|---|---|---|---|---|---|---|---|---|
| | 3.1 | | | | | | | | | | |
| | No. | % | Adj. | No. | % | Adj. | No. | % | Adj. | No. | % |
| 1: Shows solidarity | 0 | 0 | −0.06 | 1 | 0 | 0.5 | 1 | 0 | 0 | 2 | 0 |
| 2: Shows tension release | 2 | 1 | −0.5 | 2 | 0 | −1.9 | 12 | 2 | 2.1 | 16 | 1 |
| 3: Agrees | 31 | 12 | 0.9 | 73 | 14 | 3 | 58 | 8 | −3.5 | 162 | 10 |
| Total positive socio-emotional | | | | | | | | | | | 11 |
| 4: Gives suggestion | 19 | 7 | −0.7 | 37 | 7 | −1.6 | 75 | 10 | 2 | 131 | 8 |
| 5: Gives opinion | 41 | 16 | −2.4 | 102 | 19 | −1.6 | 189 | 25 | 3.4 | 332 | 21 |
| 6: Gives information | 87 | 33 | −0.8 | 209 | 39 | 2 | 258 | 34 | −1.3 | 554 | 35 |
| Total giving task-based | | | | | | | | | | | 65 |
| 7: Asks for information | 27 | 10 | 0.7 | 54 | 10 | 0.9 | 62 | 8 | −1.4 | 143 | 9 |
| 8: Asks for opinion | 18 | 7 | 0.5 | 22 | 4 | −3 | 57 | 7 | 2 | 97 | 6 |
| 9: Asks for suggestion | 11 | 4 | 2.6 | 7 | 1 | −1.6 | 15 | 2 | −0.4 | 33 | 2 |
| Total requesting task-based | | | | | | | | | | | 17 |
| 10: Disagrees | 12 | 5 | 0.4 | 31 | 6 | 2.3 | 22 | 3 | −2.5 | 65 | 4 |
| 11: Shows tension | 15 | 6 | 4.6 | 4 | 1 | −2.6 | 13 | 2 | −0.9 | 32 | 2 |
| 12: Shows antagonism | 0 | 0 | −0.9 | 0 | 0 | −1.5 | 4 | 1 | 2.1 | 4 | 0 |
| Total negative socio-emotional | | | | | | | | | | | 6 |
| Total no. | 263 | | | 542 | | | 766 | | | 1571 | |

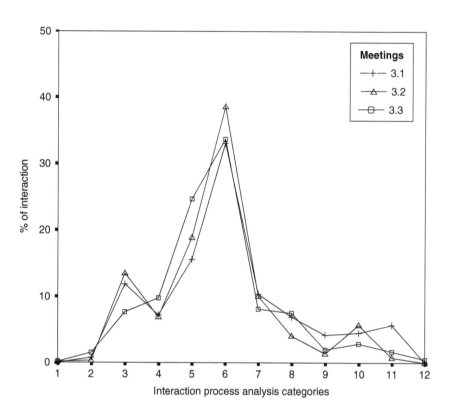

*Figure 6.3* Group interaction profiles of meetings 3.1–3.3

attempted to justify his point. Other members of the meeting were aware of the tension between these two parties and would occasionally attempt to diffuse the situation by offering information, opinion and occasionally attempting to release tension. When other members of the group realised that the discussion between the architect and the client's representative was becoming tense, they became notably quiet. Again in this case study, the disagreements between parties meant that issues were considered from different perspectives and although arguments were often tense, a course of action was often proposed and accepted by the group. Although such arguments did threaten professional relationships, they often resulted in a clear direction.

Even with this behaviour, the total amount of positive socio-emotional communication observed (11 per cent) was almost twice that of negative emotional communication (6 per cent). The amount of agreeing was relatively high, only case study one being higher; however, disagreeing was also high, only case study two having had a higher level of disagreement.

This case study had a high level of giving information, which was considerably higher than giving opinion, evaluation and analysis. The giving of information, opinions and suggestions during this series of meetings was over four times higher than asking for information, opinions and suggestions.

Tension levels were high compared with other projects, but this did not result in formal conflict or legal disputes once the contract was completed. The conflict seemed to manifest from personal tension between the architect and the client's project manager. Comments made by the client's project manager would irritate the architect. The client's project manager was young, ambitious and eager to put forward his views. The architect was the most senior member of the team and often challenged and criticised the views put forward by the project manager. Disagreements developed in all of the meetings, but it did not dominate the meeting, nor did it result in reoccurring disagreements about the same issue in subsequent meetings.

### *Summary*

Informal interaction took place prior to the meeting, but the meeting was conducted in a formal and sober manner. An agenda was circulated prior to the meeting. At the relevant item on the agenda, each sub-contractor present was asked to report on progress. The contractor's project manager would high-light relevant problems that needed resolution and suggest what type of action was required. The project manager was keen to ensure that each problem was dealt with properly, safely and within the project's time constraints. Matters were discussed in detail. Where issues were impacting on other tasks or programme slippage had occurred, the project programme was discussed and options considered. A summary of the action agreed was briefly discussed before moving on to the next item of the agenda.

## Results: Case study 4

This case study project comprised an element of refurbishment and new build work. An inner-city four-storey high building was converted and enlarged to create a number of luxury apartments. The project was managed under a general contract. The work was not completed within the budget, nor was it completed within the scheduled programme. There was evidence of conflict with the client on this project.

### *Observation of meetings 4.1–4.3*

Compared with other cases, a high level of social interaction occurred both prior to and after the progress meetings. The project manager did not get involved in the conversations prior to the meeting as he was busy reproducing (photocopying) minutes and reports for the other parties to use. The structural engineer often arrived late and after the meeting had started. The architect arrived late in meeting 4.2. This was put down to problems with finding a parking space for the car near to the city centre development. The problem of insufficient car parking spaces in close proximity to the site was one of the topics often raised in the social discussion prior to the start of the meeting.

There was considerable variation in personnel attending this series of meetings. Six people attended meetings 4.1, then ten people attended meeting 4.2 and finally seven people attended the last meeting observed. The increase in numbers during meeting 4.2 was largely attributable to an increase in the contractor's representatives. It was found that when new members entered the meeting the amount of socio-emotional interaction reduced and explanation regarding task-related issues increased slightly. As the new members, who were less familiar with the project, started to engage in discussions a greater explanation of issues was noted.

## Analysis

The three line graphs present a pattern of interaction that has a degree of consistency, although the variation between the lines is more notable than most other cases (Figure 6.4). A factor that needs to be considered when examining the degree of variation is the changing attendance and participation of contractor's representatives.

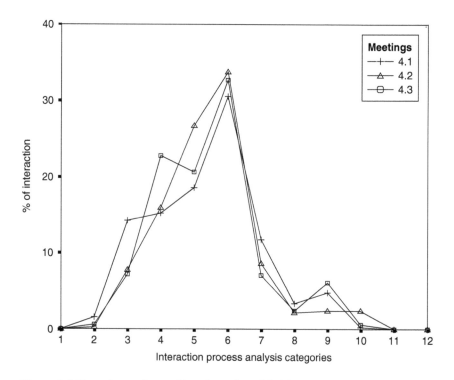

*Figure 6.4* Group interaction profiles of meetings 4.1–4.3

In this case study the most consistent categories observed were giving information and asking for opinions (Table 6.5). The greatest variation occurred in the amount of agreement. The variation between meetings associated with both giving suggestions and disagreeing result in high adjusted residuals.

The total amount of task-based communication accounts for over 88 per cent of the interaction and the vast majority of the remainder of the interaction are in the positive emotional categories. The amount of positive emotional communication is over ten times greater than the negative communication. Although this ratio is high, the amount of positive emotional communication (10 per cent) was not the highest observed. The factor that contributes the most to the high ratio is the low level of negative emotional communication (1 per cent). Showing solidarity, showing tension and showing antagonism were not observed in any of the meetings.

Very little critical discussion regarding problems occurred in these meetings. Many of the issues continued to be discussed from meeting to meeting, seemingly reoccurring without resolution. Participants had gathered and presented further information at each meeting to support the understanding of the problem, yet no one person was prominent in offering suggestions or giving direction (compared with, e.g. Case study 1). The inability to make decisions may have had an affect on the inability of the contractor to complete the project on time or within budget. Problems would be discussed and reasons for not being able to resolve the matter put forward without any suggested way forward.

### Summary

The contractor's project manager was rather unorganised. The minutes of the previous meeting were often prepared and copied immediately prior to the start of the meeting. The minutes of the meeting and the agenda were neither easy to understand nor systematically structured. Action points were recorded, but were open to interpretation.

The meeting did not follow an organised structure and was further disrupted by members entering the meeting after it had started. Although each actor contributed to the meeting, the meeting lacked organisation and failed to address key issues. Much of the communication was focused on reports and people reading through their progress reports (which did not seem to deal with all the problems that needed addressing). When problems were discussed the contractor's representative often became frustrated. The anxiety resulted in heightened moments of negative socio-emotional expressions. These outbursts were not personalised, but expressed frustration with a situation. Unfortunately, the frustration did not result in action. Other members attending the meeting remained silent until the frustration had subsided, they did not attempt to pacify the project manager. The frustration did not prompt others to take or suggest a course of action. Although the frustration was expressed in relation to a number of inter-related problems, the specific cause and person or persons considered responsible for the problems were not easily identifiable. As no specific person or situation was identified as

Table 6.5 Results of group interaction for meetings 4.1–4.3

| Interaction categories | Meeting 4.1 | | | 4.2 | | | 4.3 | | | Total | |
|---|---|---|---|---|---|---|---|---|---|---|---|
| | No. | % | Adj. | No. | % | Adj. | No. | % | Adj. | No. | % |
| 1: Shows solidarity | 0 | 0 | 0 | 0 | 0 | 0 | 0 | 0 | 0 | 0 | 0 |
| 2: Shows tension release | 8 | 2 | 2.3 | 1 | 0 | −1.6 | 4 | 1 | −0.7 | 13 | 1 |
| 3: Agrees | 75 | 14 | 4.3 | 35 | 8 | −1.6 | 47 | 7 | −2.7 | 157 | 10 |
| *Total positive socio-emotional* | | | | | | | | | | | *10* |
| 4: Gives suggestion | 80 | 15 | −2.3 | 72 | 16 | −1.6 | 148 | 23 | 3.7 | 300 | 18 |
| 5: Gives opinion | 98 | 19 | −2.1 | 121 | 27 | 3.1 | 134 | 21 | −0.9 | 353 | 22 |
| 6: Gives information | 161 | 31 | −1.1 | 153 | 34 | 0.8 | 213 | 33 | 0.3 | 527 | 32 |
| *Total giving task-based* | | | | | | | | | | | *72* |
| 7: Asks for information | 62 | 12 | 2.7 | 39 | 9 | −0.3 | 46 | 7 | −2.2 | 147 | 9 |
| 8: Asks for opinion | 18 | 3 | 1.2 | 10 | 2 | −0.8 | 16 | 3 | −0.5 | 44 | 3 |
| 9: Asks for suggestion | 25 | 5 | 0.1 | 11 | 2 | −2.6 | 40 | 6 | 2.3 | 76 | 5 |
| *Total requesting task-based* | | | | | | | | | | | *16* |
| 10: Disagrees | 1 | 0 | −2.2 | 11 | 2 | 3.7 | 4 | 1 | −1.2 | 16 | 1 |
| 11: Shows tension | 0 | 0 | 0 | 0 | 0 | 0 | 0 | 0 | 0 | 0 | 0 |
| 12: Shows antagonism | 0 | 0 | 0 | 0 | 0 | 0 | 0 | 0 | 0 | 0 | 0 |
| *Total negative socio-emotional* | | | | | | | | | | | *1* |
| *Total no.* | *528* | | | *453* | | | *652* | | | *1633* | |

Note
Adj. = Adjusted residuals – standardised difference between observed and expected.

the cause of the problem, none of the members of the M&D team were asked or prompted to respond to the outburst.

## Results: Case study 5

This design and build project comprised a two-storey medical centre, with retail units and other public amenities. The work was completed within budget and within the scheduled programme. There was no evidence of conflict.

### Observations of meetings 5.1–5.4

Discussions took place prior to the commencement of the meetings and these were normally within subgroups, that is people would generally talk with people from the same organisation. The mechanical and electrical consultant did initiate some communication across organisational boundaries. It did seem that while some people would engage in conversations, others were not as forthcoming. The contractor's representatives avoided conversations with the mechanical and electrical consultant, even though the mechanical and electrical contractor did attempt to initiate conversations.

### Analysis

The second and third meetings had a greater consistency than the first and final meeting observed (Figure 6.5 and Table 6.6). Meeting 5.4 took place towards the

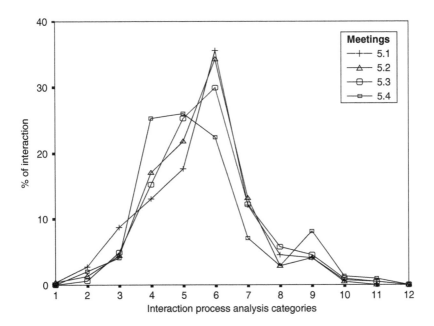

*Figure 6.5* Group interaction profiles of meetings 5.1–5.4

Table 6.6 Results of group interaction for meetings 5.1–5.4

| Interaction categories | Meeting 5.1 | | | 5.2 | | | 5.3 | | | 5.4 | | | Total | |
|---|---|---|---|---|---|---|---|---|---|---|---|---|---|---|
| | No. | % | Adj. | No. | % | Adj. | No. | % | Adj. | No. | % | Adj. | No. | % |
| 1: Shows solidarity | 2 | 0 | 1.3 | 2 | 0 | 1.5 | 0 | 0 | −1.1 | 0 | 0 | −1.4 | 4 | 0 |
| 2: Shows tension release | 18 | 3 | 2.3 | 8 | 1 | −0.7 | 4 | 1 | −2.6 | 19 | 2 | 0.9 | 49 | 2 |
| 3: Agrees | 58 | 9 | 4.3 | 26 | 5 | −1.1 | 34 | 5 | −0.7 | 38 | 4 | −2.3 | 156 | 5 |
| Total positive socio-emotional | | | | | | | | | | | | | | 7 |
| 4: Gives suggestion | 86 | 13 | −4 | 99 | 17 | −0.9 | 106 | 15 | −2.4 | 237 | 25 | 6.7 | 528 | 18 |
| 5: Gives opinion | 116 | 18 | −3.8 | 127 | 22 | −0.8 | 176 | 25 | 1.6 | 243 | 26 | 2.6 | 662 | 23 |
| 6: Gives information | 235 | 36 | 3.8 | 199 | 34 | 2.7 | 208 | 30 | 0.2 | 210 | 22 | −5.9 | 852 | 30 |
| Total giving task-based | | | | | | | | | | | | | | 71 |
| 7: Asks for information | 80 | 12 | 1.3 | 77 | 13 | 2.2 | 85 | 12 | 1.5 | 66 | 7 | −4.4 | 308 | 11 |
| 8: Asks for opinion | 30 | 5 | 0.9 | 17 | 3 | −1.4 | 40 | 6 | 2.8 | 27 | 3 | −2.1 | 114 | 4 |
| 9: Asks for suggestion | 27 | 4 | −1.8 | 24 | 4 | −1.7 | 32 | 5 | −1.2 | 76 | 8 | 4.2 | 159 | 6 |
| Total requesting task-based | | | | | | | | | | | | | | 20 |
| 10: Disagrees | 5 | 1 | −0.6 | 3 | 1 | −1.2 | 7 | 1 | 0.2 | 12 | 1 | 1.3 | 27 | 1 |
| 11: Shows tension | 3 | 1 | −0.3 | 0 | 0 | −2 | 3 | 0 | −0.4 | 9 | 1 | 2.3 | 15 | 1 |
| 12: Shows antagonism | 0 | 0 | 0 | 0 | 0 | 0 | 0 | 0 | 0 | 0 | 0 | 0 | 0 | 0 |
| Total negative socio-emotional | | | | | | | | | | | | | | 1 |
| Total no. | 660 | | | 582 | | | 695 | | | 937 | | | 2874 | |

Note

Adj. = Adjusted residuals − standardised difference between observed and expected.

end of the project, although it was not the final meeting. The results of meetings observed towards the end of the project show changes in interaction patterns.

High adjusted residuals for category 4 (giving suggestions) were found in meetings 5.4 and 5.1; this was the largest variation found in case study 5. The levels of giving opinions, giving information and asking for suggestions did vary in individual meetings, indicated by the high adjusted residuals. The categories that are most consistent between meetings were agreeing and disagreeing. Much of the variation occurred in the neutral socio-emotional categories. Giving information, opinions and suggestions accounts for six times the amount of interaction associated with requesting the same types of interaction.

The total task-based communication accounts for over 90 per cent of the total interaction. The majority of the remaining interaction is positive emotional (7 per cent); moreover, the positive emotional communication exceeds the negative emotional communication by a factor of five.

It was evident that the contractor's representatives were not satisfied with the performance of the mechanical and electrical consultant. Much of the disagreement and tension observed was attributable to the M&D of services within the building.

Occasionally, the team members did express negative emotion, although this was mainly confined within one particular dyad. There was a degree of emotionally charged debate between the mechanical and electrical consultant and the contracts' manager. The planner also entered into the discussion, tending to use occasional negative emotional communication. The contractor's representatives raised concerns over the action taken by the mechanical and electrical consultant. They clearly showed that they were not satisfied with the consultant's performance.

The conflict and the complexity of the issues being discussed between contractors and mechanical and electrical representatives resulted in the mechanical and electrical sub-contractors being asked to attend the meetings. The discussions continued to be tense; however, it was evident in the later meetings that many issues were resolved and courses of action proposed and agreed. In this situation, the negative socio-emotional interaction was followed by suggested action. Initially, the action suggested was to ensure that all the main parties who were potentially involved in the problem were present at the meeting. This ensured that no other party outside the meeting could be blamed for the problems occurring. Once all parties were present in the meeting the focus was on how the position could be rectified and the time scales involved to rectify the problem.

### Summary

An agenda was produced prior to the meeting. Issues on the agenda were briefly discussed and reported. The chair moved quickly on to the item that required a report from the mechanical and electrical consultant. The consultant was allowed to report on progress, but was quickly questioned about problems with wiring and delays with the design information. The meeting was clearly focused on attacking and resolving this problem.

Questions were directed at the consultant from the architect, quantity surveyor and project manager. Some of the questions had an accusing tone. It was clear that the mechanical and electrical consultant was considered to be the cause of the problem. The focus of the tension expressed was aimed at the mechanical and electrical consultants. During the initial meetings the mechanical and electrical consultant refused to accept responsibility for the problems and identified a number of other issues that were impacting on the problem. The mechanical and electrical consultant suggested that there were a number of issues associated with the mechanical and electrical contractor, which had impacted on the main problem and delay.

When the problem remained unresolved in previous meetings, the mechanical and electrical subcontractor was asked to join the meeting. With up to nine professionals meeting in a small site cabin (also used to store files) conditions were cramped, however the meeting was conducted in a formal manner. At times during the meeting there was tension between the members of the meeting. Tension emerged as bouts of silence, followed by formal questions aimed at the mechanical and electrical consultant. With all relevant specialist in the room, the contractor's representatives and the architect quickly dealt with the problems. It was decided that other issues on the agenda should be dealt with after the mechanical and electrical matters had been covered. It was also stated that this would allow the mechanical and electrical contractors to leave the meeting, if they wished, once the matter had been discussed. With all parties present and focus on the mechanical and electrical consultant, matters were soon resolved. The mechanical and electrical contractor made a suggestion, the architect asked the consultant whether this was possible, the consultant agreed and the matter was resolved.

## Results: Case study 6

Case study 6 comprised a new school building with parking facilities, which was managed under a general contract. The work was not completed within budget, nor was it completed within the time schedule. This was the only project where evidence of conflict with the client and with the architect was observed and was the least successful of the ten projects.

### Observation of meetings 6.1–6.4

Prior to the start of the meetings most of the professionals present in the room took part in informal conversations. The contractor's representatives did not enter the room until just before the meeting was due to start and did not engage in any discussions prior to the commencement of the meeting. Informal discussions concerning issues raised in the meeting took place after the meeting had finished.

## Analysis

Prior to the start of the observations, a change took place in the contractor's management team and a new project manager was appointed. No personnel changes were made during the observation period. This meant that the contractor's planner took on the role of project manager for the first meeting. During the first meeting the contracts' manager and planner led the interactions for the contractor's representatives. The new project manager, who at this point was unfamiliar with both the project and the participants, did not make major contributions until the second meeting, where he was a major contributor. His contributions increased at each subsequent meeting. Although the architect chaired the meeting, he was not dominant in directing interaction. Other meeting participants (mainly the contractor's representatives) took on this role and influenced much of the interaction.

All of the line graphs (Figure 6.6) of case study 6 follow a similar profile. The greatest variation in the profile occurs in the task-based categories (Table 6.7). The most consistent interaction results were found in the agreement category. The categories with the highest adjusted residual, indicating greatest variation between meetings, include showing tension, giving information and asking for information. Over the course of the meetings the amount of tension increased, which corresponds with the pressure from the contractor's representatives to resolve

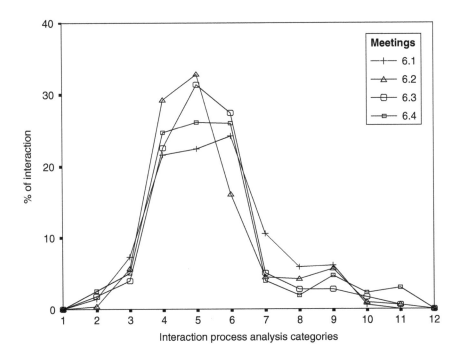

*Figure 6.6* Group interaction profiles of meeting 6.1–6.4

Table 6.7 Results of group interaction for meetings 6.1–6.4

| Interaction categories | Meeting 6.1 | | | 6.2 | | | 6.3 | | | 6.4 | | | Total | |
|---|---|---|---|---|---|---|---|---|---|---|---|---|---|---|
| | No. | % | Adj. | No. | % | Adj. | No. | % | Adj. | No. | % | Adj. | No. | % |
| 1: Shows solidarity | 0 | 0 | 0 | 0 | 0 | 0 | 0 | 0 | 0 | 0 | 0 | 0 | 0 | 0 |
| 2: Shows tension release | 5 | 2 | −0.4 | 2 | 0 | −2.6 | 11 | 2 | 0.1 | 24 | 3 | 2.3 | 42 | 2 |
| 3: Agrees | 25 | 7 | 1.9 | 28 | 6 | 0.5 | 24 | 4 | −1.6 | 49 | 5 | −0.3 | 126 | 5 |
| Total positive socio-emotional | | | | | | | | | | | | | | 7 |
| 4: Gives suggestion | 74 | 22 | −1.4 | 146 | 29 | 2.7 | 138 | 23 | −1.4 | 241 | 25 | 0.1 | 599 | 25 |
| 5: Gives opinion | 77 | 22 | −2.6 | 164 | 33 | 2.5 | 192 | 31 | 1.9 | 255 | 26 | −1.9 | 688 | 28 |
| 6: Gives information | 83 | 24 | 0.1 | 80 | 16 | −4.7 | 168 | 28 | 2.3 | 253 | 26 | 1.8 | 584 | 24 |
| Total giving task-based | | | | | | | | | | | | | | 77 |
| 7: Asks for information | 36 | 11 | 4.7 | 22 | 4 | −0.9 | 31 | 5 | −0.2 | 38 | 4 | −2.4 | 127 | 5 |
| 8: Asks for opinion | 20 | 6 | 3 | 21 | 4 | 1.5 | 17 | 3 | −0.6 | 19 | 2 | −2.8 | 77 | 3 |
| 9: Asks for suggestion | 21 | 6 | 1.5 | 28 | 6 | 1.2 | 17 | 3 | −2.5 | 45 | 5 | 0.1 | 111 | 5 |
| Total requesting task-based | | | | | | | | | | | | | | 13 |
| 10: Disagrees | 2 | 0.6 | −1.6 | 5 | 1 | −1.2 | 10 | 2 | 0.1 | 22 | 2 | 2.1 | 39 | 2 |
| 11: Shows tension | 0 | 0 | −2.5 | 3 | 1 | −1.8 | 4 | 1 | −2.0 | 29 | 3 | 5 | 36 | 2 |
| 12: Shows antagonism | 0 | 0 | 0 | 0 | 0 | 0 | 0 | 0 | 0 | 0 | 0 | 0 | 0 | 0 |
| Total negative socio-emotional | | | | | | | | | | | | | | 3 |
| Total no. | 343 | | | 499 | | | 612 | | | 975 | | | 2429 | |

problems. As the new project manager gained confidence and understanding of the project needs, he shifted his attention towards problems and tension manifested. While stating the nature of the problem, specific questions and requests were made to members of the design team, the phrasing of the problem implied responsibility. Regardless of whether responsibility was accepted or denied, a solution was requested or suggested.

The total task-based communication accounts for over 88 per cent of the interaction. Positive emotional interaction is over twice the amount of negative interaction, with 7 per cent being positive and 3 per cent of communication acts being negative.

### *Summary*

The change in personnel towards the end of the project has complicated the analysis of this group against the project outcomes. The new project manager was relatively inactive during the first meeting (he listened and asked questions). At subsequent meetings his interaction increased. Initially he attempted to understand the nature of the problems being discussed, at subsequent meetings his emphasis was on looking for suggestions on how the problems could be resolved and giving suggestions. The change in personnel resulted in an interpersonal and group development process; those coming into the team had to understand the problems, their context and develop relationships with the other members of the M&D team. Rather quickly, over the short period of a few meetings, those new to the project developed and changed their interaction style. Their interaction changed from information seeking, questioning and listening behaviours to an interaction which probed at comments, asked questions, then encouraged participants to make suggestions. Where suggestions were not forthcoming, directions were provided. Early interactions from the contractor's representative were rather passive; whereas later on in the project their interaction and influence increased contributing greatly to the direction of the group.

It is not possible to say whether or not the interaction of this group was representative of a failing group or a team that had managed to mitigate some of the losses of the previous group. There were a plethora of problems at the start of the observation, which were resolved as the new project team developed and took action over the course of the meetings observed.

As new members entered the group, interaction changed as the newcomers developed relationships with the project team. These relationships were for the benefit of the project, and would not be described as friendships. Newcomers used their interaction to gather information on the project and those responsible for areas of work; the project problems and those responsible or associated with the problem. By interacting with the members of the team, the new members soon developed an ability to interact with each member, assimilating the information required and gaining the desired cooperation to overcome problems. Most of the interaction was formal, occasional negative emotion and tension emerged when solutions to problems were not forthcoming, although some 'light-hearted' positive emotion was also used, which did reduce tension in the meeting.

## Results: Case study 7

This project involved elements of refurbishment and new-build. The development was a five-storey building that contained a restaurant and luxury apartments. Managed under a general contract the work was not completed within budget, nor was it completed within the scheduled timeframe. There was evidence of conflict with the client.

### Observations of meetings 7.1–7.4

During the earlier meetings, informal conversations concerning the building often took place prior to the meeting; however, in the later meetings such socialising activity was rarely observed. As problems developed, informal interaction (social or 'small talk') did not take place. Conversations that took place were quiet and discrete. It appeared that the participants involved in these discussions did not want other people to hear their conversation.

### Analysis

Although procured under a traditional contract, with the architect responsible for administration on behalf of the client, the architect did not actively chair the meetings. The contractor's representatives played a much greater role in directing the meetings. The first two meetings took place in what was considered a normal atmosphere, while the last two meetings observed were tense.

All four profiles follow a pattern of interaction that has a degree of consistency across all of the meetings (Figure 6.7). The greatest variation occurs in meeting 7.1 agreeing, where an adjusted residual of 7.1 for 'agrees' is found (Table 6.8). Other high adjusted residuals include meeting 7.2 requesting suggestions, meeting 7.4 showing tension and meeting 7.3 giving information. The amount of agreeing is much higher in the first meeting than in any of the subsequent meetings, and the amount of tension (category 11) is considerably higher in the last meeting observed than in any of the earlier meetings. As the amount of interaction associated with giving information increased, the amount of suggestions made was reduced. Giving information, opinions and suggestions is almost five times the value of requests for information, opinions and suggestions.

The amount of task-based interaction accounts for over 92 per cent of the total interaction. The positive emotional communication accounts for 6 per cent of the interaction and negative socio-emotion communication represents 2 per cent of the total interaction.

During the later meetings, the contractor's project manager generated much of the negative emotional interaction. The negative emotional expression used by the project manager seemed to be an expression of frustration related to past events; it was difficult to identify whom it was aimed at, and whether a response was required. When the project manager used negative emotional interaction the other professionals seemed unsure how to react. In the vast majority of the cases, when

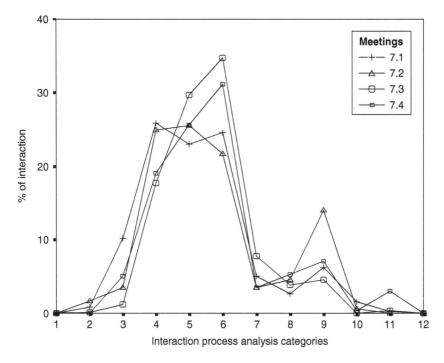

*Figure 6.7* Group interaction profiles of meetings 7.1–7.4

the project manager showed such emotion, the other professionals expressed no direct verbal response. The meeting changed from an open information exchange to a meeting where participants became very defensive.

The negative emotional communication in this case study is indicated by disagreements and showing tension. This is the only situation where the amount of tension exceeds the disagreements. Most of the tension that was observed in this case study was attributable to the frustration shown by the project manager. The frustration shown in the last meeting accounted for 3 per cent of the total interaction. This is almost twice the amount of tension found in any other meeting observed during this research. Although many different communication acts can be classified as 'showing tension', generally this category was used to express anger with something or someone. The project manager often expressed general frustration with large areas of the project, making it difficult to point to the exact nature of the problem or the person he considered responsible for the problem.

Positive socio-emotional interaction is almost four times greater than the negative emotional interaction. The amount of socio-emotional interaction in meeting 7.2 and 7.3 is low. There is a notable increase in the amount of agreeing shown in meeting 7.4; much of this was in reaction to the negative emotional expression. Because the exact target of the negative emotional interaction was

Table 6.8 Results of group interaction for meetings 7.1–7.2

| Interaction categories | Meeting | | | | | | | | | | | | Total | |
|---|---|---|---|---|---|---|---|---|---|---|---|---|---|---|
| | 7.1 | | | 7.2 | | | 7.3 | | | 7.4 | | | | |
| | No. | % | Adj. | No. | % | Adj. | No. | % | Adj. | No. | % | Adj. | No. | % |
| 1: Shows solidarity | 0 | 0 | -0.7 | 0 | 0 | -0.5 | 0 | 0 | -0.5 | 1 | 0 | 1.8 | 1 | 0 |
| 2: Shows tension release | 7 | 1 | 0.8 | 8 | 2 | 3.1 | 1 | 0 | -1.6 | 0 | 0 | -2.3 | 16 | 1 |
| 3: Agrees | 87 | 10 | 7.1 | 17 | 4 | -2.3 | 7 | 1 | -5.2 | 31 | 5 | -0.7 | 142 | 6 |
| *Total positive socio-emotional* | | | | | | | | | | | | | | *6* |
| 4: Gives suggestion | 221 | 26 | 3.2 | 121 | 25 | 1.6 | 102 | 18 | -2.9 | 116 | 19 | -2.2 | 560 | 22 |
| 5: Gives opinion | 196 | 23 | -2.2 | 124 | 26 | 0 | 171 | 30 | 2.5 | 156 | 26 | 0 | 647 | 26 |
| 6: Gives information | 209 | 25 | -2.7 | 105 | 22 | -3.4 | 200 | 35 | 4.2 | 190 | 31 | 2.1 | 704 | 28 |
| *Total giving task-based* | | | | | | | | | | | | | | *76* |
| 7: Asks for information | 43 | 5 | 0.1 | 17 | 4 | -1.7 | 45 | 8 | 3.5 | 21 | 3 | -2.0 | 126 | 5 |
| 8: Asks for opinion | 22 | 3 | -2.4 | 22 | 5 | 0.8 | 22 | 4 | -0.1 | 32 | 5 | 2 | 98 | 4 |
| 9: Asks for suggestion | 53 | 6 | -1.8 | 68 | 14 | 6 | 26 | 5 | -3.1 | 43 | 7 | -0.5 | 190 | 8 |
| *Total requesting task-based* | | | | | | | | | | | | | | *16* |
| 10: Disagrees | 13 | 2 | 3.5 | 3 | 1 | -0.3 | 0 | 0 | -2.3 | 2 | 0 | -1.3 | 18 | 1 |
| 11: Shows tension | 2 | 0 | -2.5 | 0 | 0 | -2.3 | 2 | 0 | -1.5 | 18 | 3 | 6.3 | 22 | 1 |
| 12: Shows antagonism | 0 | 0 | 0 | 0 | 0 | 0 | 0 | 0 | 0 | 0 | 0 | 0 | 0 | 0 |
| *Total negative socio-emotional* | | | | | | | | | | | | | | *2* |
| *Total no.* | 853 | | | 485 | | | 576 | | | 610 | | | 2524 | |

not clear, the resulting expressions of agreement on such issues was also vague. The socio-emotional interactions in this case study were more erratic than the other cases.

### *Summary*

Task-based interaction in this case study was high. When discussing progress or issues there was a tendency to exchange information that was general in nature, providing background information, identifying events, recounting occurrences and conversations. The focus was on information and events surrounding problems rather than identifying the nature of the problem and trying to resolving it. Due to the amount of general information put forward, discussions seemed prolonged and members appeared to lose interest and became distracted. Over the course of the meeting the amount of suggestions provided to resolve problems reduced. The tension expressed coincided with a commensurate fall in suggestions to problems.

## Results: Case study 8

This case study project was managed under a general contract and comprised the refurbishment of a four-storey building to provide a restaurant unit and apartments. The contractor completed the work within budget, but not within the scheduled programme. There was no evidence of conflict on this project.

### *Observation of meetings 8.1–8.4*

All of the participants, other than the contractor's representatives, arrived on time or just before the meetings were about to start. Due to the participants arriving just before the meeting, there was little time for socialising. In preparation for the meetings, the contracts' manager and the project manager would often discuss problems, any action taken or action considered necessary to resolve the problems, prior to the commencement of the meeting. The quantity surveyor was normally present in the room with the project manager and contract manager during these discussions, but would only intervene on issues of cost and contract.

### *Analysis*

The four line graphs show relatively consistent levels of interaction between meetings (Figure 6.8). The lines skew slightly to the left hand side of the task dimensions, having higher levels of interaction associated with giving information, opinions and suggestions. The level of information giving in three of the meetings is lower than the giving of opinions or suggestions (Table 6.9).

The amount of task-based interaction, in Case study 8, accounts for over 92 per cent of the total interaction; the majority of the remaining of interaction is positive emotional. The amount of positive emotional communication is almost three times greater than the negative emotional communication. Extremes of

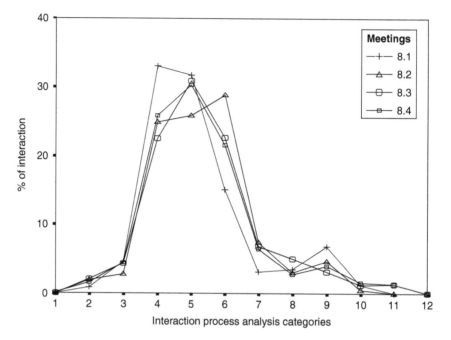

*Figure 6.8* Group interaction profiles of meetings 8.1–8.4

negative emotions were not used, showing antagonism was not observed in the meetings associated with this case study.

A client's project manager attended all of the meetings and it was clear from the interaction observed that the architect afforded him considerable respect. During the meetings the architect was very attentive to the client's expressions and reactions, being quick to offer explanation to the points raised. Although not disrespectful, the other professionals did not extend the same respect or attention to the client's representative.

### Summary

Prior to the meetings, the client's team (project manager, quantity surveyor and architect) would walk around the site inspecting the work and assessing progress; it was clear that there was some dialogue conducted on site prior to the meeting, to gauge whether problems were in hand. Occasionally, the client's team would consult with the contractor's site manager regarding progress and problems, before the meeting commenced. The meetings were conducted in a formal manner. Participants were asked to report on progress of their aspect of work. When progress was as expected, comments and reports were fairly short and the client's project manager was quick to move on to the problems. Discussions regarding potential problems were dealt with in a firm and formal manner. The client's project

Table 6.9 Results of group interaction for meetings 8.1–8.4

| Interaction categories | Meeting | | | | | | | | | | | | Total | |
| --- | --- | --- | --- | --- | --- | --- | --- | --- | --- | --- | --- | --- | --- | --- |
| | 8.1 | | | 8.2 | | | 8.3 | | | 8.4 | | | | |
| | No. | % | Adj. | No. | % | Adj. | No. | % | Adj. | No. | % | Adj. | No. | % |
| 1: Shows solidarity | 0 | 0 | −0.6 | 0 | 0 | −0.5 | 0 | 0 | −0.6 | 1 | 0 | 1.5 | 1 | 0 |
| 2: Shows tension release | 5 | 1 | −1.6 | 8 | 2 | 0.7 | 12 | 2 | 1 | 11 | 2 | 0 | 36 | 2 |
| 3: Agrees | 26 | 5 | 0.6 | 11 | 3 | −1.6 | 25 | 4 | 0.2 | 31 | 5 | 0.6 | 93 | 4 |
| *Total positive socio-emotional* | | | | | | | | | | | | | | *6* |
| 4: Gives suggestion | 189 | 33 | 4 | 101 | 25 | −0.9 | 131 | 23 | −2.6 | 179 | 26 | −0.6 | 600 | 27 |
| 5: Gives opinion | 181 | 32 | 1 | 105 | 26 | −2.0 | 180 | 31 | 0.6 | 209 | 30 | 0.2 | 675 | 30 |
| 6: Gives information | 86 | 15 | −4.3 | 117 | 29 | 4 | 132 | 23 | 0.8 | 149 | 22 | 0 | 484 | 22 |
| *Total giving task-based* | | | | | | | | | | | | | | *79* |
| 7: Asks for information | 18 | 3 | −3.2 | 30 | 7 | 1.4 | 40 | 7 | 1.1 | 45 | 7 | 0.8 | 133 | 6 |
| 8: Asks for opinion | 20 | 4 | −0.1 | 12 | 3 | −0.7 | 29 | 5 | 2.2 | 19 | 3 | −1.4 | 80 | 4 |
| 9: Asks for suggestion | 39 | 7 | 3 | 19 | 5 | 0.1 | 18 | 3 | −2.0 | 27 | 4 | −1 | 103 | 5 |
| *Total requesting task-based* | | | | | | | | | | | | | | *14* |
| 10: Disagrees | 7 | 1 | 0.1 | 2 | 1 | −1.4 | 7 | 1 | 0 | 11 | 2 | 1.1 | 27 | 1 |
| 11: Shows tension | 0 | 0 | −2.4 | 0 | 0 | −1.9 | 8 | 1 | 2 | 9 | 1 | 2 | 17 | 1 |
| 12: Shows antagonism | 0 | 0 | 0 | 0 | 0 | 0 | 0 | 0 | 0 | 0 | 0 | 0 | 0 | 0 |
| *Total negative socio-emotional* | | | | | | | | | | | | | | *2* |
| Total no. | 571 | | | 405 | | | 582 | | | 691 | | | 2249 | |

Note
Adj. = Adjusted residuals − standardised difference between observed and expected.

manager placed considerable emphasis on the importance of adequate resolution. Where problems had occurred and/or he anticipated difficulties associated with planned activities, his manner showed that he expected to be reassured that events were under control. Most of the meetings were conducted in a very formal environment, but when problems had been discussed and reassurance given that they were now in hand, other members of the M&D team would occasionally intervene with 'light hearted' statements or a joking comment that would help relieve some of the tension that had developed. Taking an overview of the group atmosphere, the meeting could be split between very formal 'professional but firm' interaction that dealt with issues that were clearly given a high priority, to short episodes of more relaxed interaction. Statements that relieved tension did not come from the client's project manager. The client's project manager, who upheld a formal and business-like manner throughout the meeting, did not attempt to stop informal interaction, neither did he openly engage in jokes or light-hearted statements.

This project had the highest number of personnel attending the meetings. Although the meetings were conducted in a large room, which was part of the building that was being refurbished, the table arrangement was rather awkward and untidy. Tables that were of slightly different heights and shapes were pushed together to make one large table arrangement. Although the table arrangement was slightly untidy this did not seem to have a negative effect on the meeting. The contracting team hosted the meeting; however, the architect led and chaired the meeting.

## Results: Case study 9

Case study 9 was the refurbishment and extension of a hospital wing, which was managed under a general contract. The work was successfully completed within budget and to programme. There was no evidence of conflict. The refurbishment of the hospital facility was completed in a number of phases while the hospital facilities and wards adjacent to building works continued to be used. To ensure that patients and hospital employees were not put at risk and so that hospital activities could continue, the construction operations had to be coordinated with the hospital management.

### Observations of meetings 9.1–9.3

Informal discussions did take place between various people prior to the meetings. The majority of the professionals were friendly, making an effort to greet people as they met and sat down at the meeting table. Drinks were offered to all who attended the meeting.

### Analysis

These meetings were quite different from the other case studies, due to the number of non-construction professionals that represented the client. In addition to the professional consultants employed by the client, it was normal for at least three representatives from the client organisation, a health trust, to be present in the meetings.

The architect chaired the meeting and ensured that the participants' contributions were focused on the agenda. Occasionally he allowed conversations to drift away from the point for a few seconds but would then politely and firmly ask the speaker to return to the issues. The general atmosphere of the meeting was relaxed and professional.

The three line graphs present a consistent set of profiles (Figure 6.9). Meeting 9.3 has a lower level of giving opinion, evaluation and analysis, and higher level of giving information than the other two meetings (Table 6.10). The profiles have high levels of interaction associated with opinion giving and information giving, and lower levels of giving suggestions.

The amount of task-based communication accounts for over 94 per cent of the interaction, 4 per cent of the interaction is positive socio-emotional and 1 per cent is negative emotional communication. While the overt disagreements

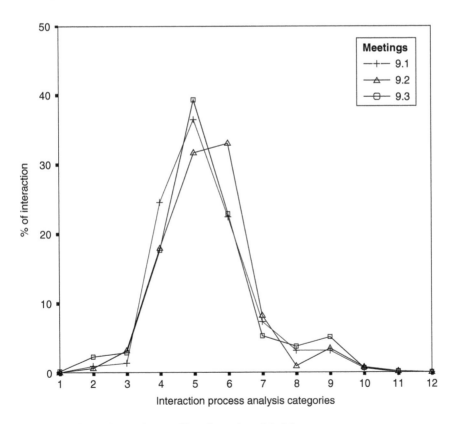

*Figure 6.9* Group interaction profiles of meetings 9.1–9.3

Table 6.10 Results of group interaction for meetings 9.1–9.3

| Interaction categories | Meeting | | | | | | | | | Total | |
|---|---|---|---|---|---|---|---|---|---|---|---|
| | 9.1 | | | 9.2 | | | 9.3 | | | | |
| | No. | % | Adj. | No. | % | Adj. | No. | % | Adj. | No. | % |
| 1: Shows solidarity | 0 | 0 | −0.6 | 0 | 0 | −0.7 | 1 | 0 | 1.2 | 1 | 0 |
| 2: Shows tension release | 4 | 1 | −0.9 | 3 | 1 | −1.8 | 14 | 2 | 2.6 | 21 | 1 |
| 3: Agrees | 6 | 1 | −1.9 | 17 | 3 | 1.2 | 18 | 3 | 0.6 | 41 | 3 |
| *Total positive socio-emotional* | | | | | | | | | | | *4* |
| 4: Gives suggestion | 110 | 25 | 3.1 | 95 | 18 | −1.2 | 111 | 18 | −1.6 | 316 | 20 |
| 5: Gives opinion | 164 | 37 | 0.2 | 168 | 32 | −2.5 | 248 | 39 | 2.2 | 580 | 36 |
| 6: Gives information | 101 | 23 | −2.1 | 175 | 33 | 4.4 | 144 | 23 | −2.4 | 420 | 26 |
| *Total giving task-based* | | | | | | | | | | | *82* |
| 7: Asks for information | 33 | 7 | 0.5 | 44 | 8 | 1.6 | 33 | 5 | −2 | 110 | 7 |
| 8: Asks for opinion | 14 | 3 | 0.7 | 5 | 1 | −3 | 24 | 4 | 2.3 | 43 | 3 |
| 9: Asks for suggestion | 14 | 3 | −1.1 | 18 | 3 | −0.8 | 32 | 5 | 1.8 | 64 | 4 |
| *Total requesting task-based* | | | | | | | | | | | *14* |
| 10: Disagrees | 3 | 1 | 0 | 4 | 1 | 0.2 | 4 | 1 | −0.2 | 11 | 1 |
| 11: Shows tension | 0 | 0 | −0.9 | 1 | 0 | 0.5 | 1 | 0 | 0.3 | 2 | 0 |
| 12: Shows antagonism | 0 | 0 | 0 | 0 | 0 | 0 | 0 | 0 | 0 | 0 | 0 |
| *Total negative socio-emotional* | | | | | | | | | | | *1* |
| Total no. | 449 | | | 530 | | | 630 | | | *1609* | |

were low, many of the client's representatives were very inquisitive of the issues raised and suggestions made. While not often disagreeing with the proposals, the client's representatives enquired about alternative methods, asked for greater explanation and offered alternative suggestions. Problems of construction and design were discussed, as were issues of how and when construction activities would take place, the impact that this would have on the management of the hospital, aspects of hospital health and safety and the impact on the patients. The method and management of construction operations had to tie in with the hospital's management processes and systems. Thus, many problems were discussed in considerable depth before action was agreed.

## Summary

Although the level of disagreement was not high, hospital staff were very clear and insistent on what construction operations could take place and when. As the contracting team discussed their proposed operations, client representatives and hospital users would put forward their concerns or reasons why operations could not take place. Through discussions the users (ward sisters and managers) developed an understanding of the construction tasks. Even though they were not construction experts, they enquired into aspects of construction and made suggestions on how tasks could be accomplished, albeit different to that planned and anticipated by the contractor. It was clearly difficult to co-ordinate construction operations and continue to run the hospital facility; however, the level of negative socio-emotional interaction was low. The hospital staff were committed to ensuring that the hospital could continue to operate safely, and were also keen to help the contractor overcome problems and progress with the works.

The client team put a considerable amount of information and opinions forward. At the end of a contribution, others who had not commented were also asked what they thought on the point being discussed. The focus of the discussion was to identify the barriers and then suggesting ideas or possible solutions to the problem. One of the ward sisters was particularly prominent and influential in the discussion. She tended to identify what could and could not happen, and encouraged other hospital staff to contribute so that they could identify ways that the contractor could be assisted in their operations. The sister was particularly good at encouraging the hospital's facilities manager to contribute to the discussion. The facilities manager contributed little apart from when questions were directed to him, yet he seemed to be a key link in ensuring that operations could take place. It was clear that the running of the hospital caused problems for the contractor, but through the ward sister, and other hospital users, options were explored so that construction could take place. The meeting was conducted in a formal manner; however, the level of agreement and light-hearted statement were high.

This was the only project where females participated in the discussion. Although we cannot say how this influenced group interaction, the positive socio-emotional interaction that emerged often resulted from interaction between female and male members. Positive emotion, humour and supportive statements with light-hearted emotional content were products of mixed gender interaction. Little negative socio-emotional interaction was observed in this group. Groups with female members tend to have a higher level of socio-emotional interaction than those with only males, but this was not observed here. The level of socio-emotional interaction observed in this meeting was lower than most other cases. The small level of positive socio-emotional content was linked to female interaction.

The female members of the group asked a lot of questions and also encouraged other members to contribute. Requests for help, support and assistance were used to draw other key members of the group into discussions to ensure operations could continue. When a decision needed to be made and action taken, the ward sister was keen to ensure the right people contributed to the task. Others who were not major participants were asked for their opinion or help. During one observation, while making a request to a rather reluctant group member, a female member made comments to the rest of the group that had the potential to raise the status of the person at whom the request was aimed. The polite 'status raising' gesture and supporting comments had the desired effect; the quiet and rather sullen member of the group made a contribution and proposed action that he would take.

## Results: Case study 10

This case study project comprised the refurbishment of a five-storey building to create a recording studio and a media centre. The project was managed under a general contract and was completed within budget and within the scheduled programme. There was no evidence of conflict on this project.

### Observation of meetings 10.1–10.4

All of the meetings were conducted in a formal manner with a meeting agenda issued at each meeting. During the early meetings there was a certain amount of tension. Prior to the meeting the client's project manager and consultants would sit in the meeting room socialising and discussing problems and possible action in preparation for the meeting. The contractor's representatives would sit in the adjacent offices doing the same activities as the client's team. During different observations the researcher had the opportunity to be in both the meeting room and adjacent accommodation.

### Analysis

The four profiles present consistent patterns of interaction (Figure 6.10). It was noted that, as the meetings progressed from 10.1 to 10.4, the amount of interaction associated with giving suggestions increased (Table 6.11).

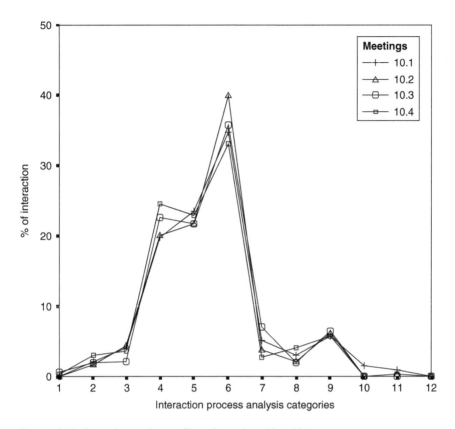

*Figure 6.10* Group interaction profiles of meetings 10.1–10.4

The total amount of task-based interaction accounts for 93 per cent of the interaction, the remainder comprises 6 per cent positive and 1 per cent negative socio-emotional communication. High levels of positive socio-emotional communication tend to be associated with successful projects. The amount of positive emotional communication is over seven times the amount of negative emotional communication.

Although negative interaction was low, as discussions delved into the specifics of problems and the contractual responsibility for tasks, the level of negative interaction would increase. As action or resolution to the problem was discussed, positive emotional responses would be used to indicate agreement.

By exposing problems in early meetings and clearly stating where members did not agree, the group was able to confront differences. In later meetings most of the differences were resolved and the group members did not engage in conflict. Meeting 10.1 was the only meeting where levels of disagreement were found, however the researcher attended two earlier meetings (interaction

Table 6.11 Results of group interaction for meetings 10.1–10.4

| Interaction categories | Meeting | | | | | | | | | | | | Total | |
|---|---|---|---|---|---|---|---|---|---|---|---|---|---|---|
| | 10.1 | | | 10.2 | | | 10.3 | | | 10.4 | | | | |
| | No. | % | Adj. | No. | % | Adj. | No. | % | Adj. | No. | % | Adj. | No. | % |
| 1: Shows solidarity | 0 | 0 | −1.3 | 0 | 0 | −0.9 | 3 | 1 | 2.1 | 1 | 0 | −0.1 | 4 | 0 |
| 2: Shows tension release | 10 | 2 | −0.2 | 5 | 2 | −0.6 | 9 | 2 | −0.5 | 13 | 3 | 1.2 | 37 | 2 |
| 3: Agrees | 19 | 4 | 0.8 | 13 | 4 | 1 | 10 | 2 | −1.8 | 16 | 4 | 0.2 | 58 | 3 |
| *Total positive socio-emotional* | | | | | | | | | | | | | | *6* |
| 4: Gives suggestion | 94 | 20 | −1.3 | 59 | 20 | −0.8 | 106 | 23 | 0.4 | 109 | 25 | 1.6 | 368 | 22 |
| 5: Gives opinion | 112 | 24 | 0.6 | 64 | 22 | −0.4 | 102 | 22 | −0.5 | 102 | 23 | 0.2 | 380 | 23 |
| 6: Gives information | 165 | 35 | −0.4 | 118 | 40 | 1.8 | 168 | 36 | 0.1 | 147 | 33 | −1.2 | 598 | 36 |
| *Total giving task-based* | | | | | | | | | | | | | | *80* |
| 7: Asks for information | 24 | 5 | 0.4 | 11 | 4 | −0.9 | 33 | 7 | 2.7 | 12 | 3 | −2.4 | 80 | 5 |
| 8: Asks for opinion | 14 | 3 | 0.2 | 6 | 2 | −0.9 | 9 | 2 | −1.4 | 18 | 4 | 1.9 | 47 | 3 |
| 9: Asks for suggestion | 27 | 6 | −0.3 | 18 | 6 | 0.1 | 30 | 6 | 0.5 | 25 | 6 | −0.3 | 100 | 6 |
| *Total requesting task-based* | | | | | | | | | | | | | | *13* |
| 10: Disagrees | 7 | 2 | 4.2 | 0 | 0 | −1.2 | 0 | 0 | −1.6 | 0 | 0 | −1.6 | 7 | 0 |
| 11: Shows tension | 4 | 1 | 2.1 | 1 | 0 | −0.1 | 0 | 0 | −1.5 | 1 | 0 | −0.5 | 6 | 0 |
| 12: Shows antagonism | 0 | 0 | 0 | 0 | 0 | 0 | 0 | 0 | 0 | 0 | 0 | 0 | 0 | 0 |
| *Total negative socio-emotional interaction* | | | | | | | | | | | | | | *1* |
| Total no. | 476 | | | 295 | | | 470 | | | 444 | | | 1685 | |

was not recorded using the IPA) in which conflict between the participants was observed.

## *Summary*

A formal agenda and minutes of the previous meeting were distributed prior to the meeting. The agenda was used to structure the meeting. When progress was on track, little time was spent discussing or reporting on the agenda item. When potential difficulties and problems experienced were tabled, the meeting became more intense, and the focus was on the problem, and responsibility; the identification of action to resolve issues brought with it some negative tension and emotion. During discussions the client's project manager, contractor and structural engineer would exchange firm and formal statements about the events and problems. The client's project manager was keen to identify responsibility and action, questions were used to both extract background information and quickly identify what action the contractor and structural engineer would undertake to overcome the problem.

During early meetings there was a clear focus on problems and ensuring that these were resolved. Matters that were considered to be under control were reported with some haste, allowing time for problems to be discussed until all associated matters were identified, a suitable resolution was suggested and responsibility for action agreed. As the number of problems reduced, the meeting became more relaxed; however, matters were still dealt with in a formal manner.

## Summary of observations

The qualitative analysis of the results found that conflict may be useful in exposing alternative perspectives to an identified problem. Disagreeing with other actors forced subjects to defend or better explain their ideas. To ensure that difficult matters were given appropriate consideration and a form of resolution agreed, dominant actors would often resort to assertive questioning. If information or resolution was not forthcoming, tension within the discussion would increase and disagreements occur. Assertive interaction that carried a level of negative emotional tension helped to elevate the importance of the discussion and often resulted in proposed solutions. Discussions were assertive and tense, but at no time did any of the group members observed demonstrate outwardly aggressive behaviour.

Meetings that were structured and conducted in a formal manner corresponded with those that were more successful. Meeting minutes, action points and agenda were used to good effect where the items were structured and clear. The length and detail of discussion contained within the minutes had little bearing on their use within the meeting. Minutes and agendas had greatest effect when they clearly and concisely identified the item reported, the group or persons responsible for providing information or action. In some cases the minutes were used to agree action, demonstrate what was agreed at the earlier meetings and identify programme slippage. This method of reporting was used to greatest effect where the party responsible suggested the action and timescale, rather than timescales being imposed by the chair or project manager. When an actor subsequently failed to deliver, they were offered the opportunity to suggest a new date; again responsibility for suggesting the date was clearly with the person responsible for the action. Repeatedly failing to honour proposals resulted in minutes that showed an embarrassing set of dates all suggested by the person at fault. The ability of the chair to request proposals and allow parties to provide their delivery time seemed particularly important. The pressure of surrounding events seemed to be enough to encourage the parties to act quickly, set timely and realistic dates, while placing the responsibility for suggestions firmly with those who had the expertise and power to undertake the action. Minutes proved to be influential where matters were clearly reported. Where dates, responsibility and action were agreed, clearly recorded and the team's progress monitored problems were successfully resolved. Minutes that were particularly detailed, attempting to include all matters discussed, often seemed to lack focus and be confusing, and were of little use during the meeting and they were too lengthy to read or follow.

While high levels of suggestion and direction in the group did not correspond with projects that were successful, where those most prominent in-group inter-action did not ask for suggestions, or give direction, many issues tabled at the meeting remained unresolved and would reoccur in subsequent meetings. When members made a suggestion and then offered direction to other parties, this resulted in the key actors confronting the problem. Even in the cases of disagree-ment, following critical and defensive discussion a course of action would often emerge. A lack of direction and firm suggestions from those most active in discussions resulted in groups discussing the same issues at subsequent meetings.

The use of positive emotions and joking often eased some of the tension that occurred during the meetings. When positive emotion followed conflict, those directly engaged in the disagreement did not immediately show the same amount of tension release as others. The positive emotion of those previously engaged in conflict developed later. Thus, those not directly engaged in conflict took the initiative to reduce the tension.

Meetings associated with unsuccessful projects did not adequately deal with problems. Time was wasted on matters that were not causing problems. Elements that were progressing as planned were reported in the same detail and sometimes in more detail than those that were causing problems. In successful projects, the focus of the meeting was aimed at any issues that had the potential to cause problems, little time was given to matters that were under control. The success of the project was related to the ability to identify aspects of work that were causing problems, or posed potential problems and deal with these properly, ensuring their potential impact on the project was controlled.

In two cases where problems emerged, the project managers became frustrated and emotional, but did not address their concern with sufficient direction to make it clear who was responsible. Rather than being assertive the project managers became irritated and demonstrated frustration. Those engaged in the meeting made no real attempt to alleviate the frustration or offer solutions to the problem.

The minutes in one of the unsuccessful cases were so poorly structured that it was difficult to follow the logic of what had been reported and who was responsible. Each item on the agenda had been given a code, the codes were supposed to follow a logical structure so that each item had a unique number following a logical sequence that could be traced back through the minutes. The coding was clumsy, numbers for different issues were repeated, mistakes were made and the numeric system proved difficult and impossible to follow. Matters reported lacked clarity, it was difficult to identify the issue, action and responsibility.

# 7   Team interaction characteristics

The data from all projects were aggregated to provide a single profile for the M&D team meeting (Figure 7.2), this was then compared against previous studies. The line graph provides a profile that represents the combined interaction of all the meetings. With few earlier studies investigating *bona fide* meetings, this provides the most comprehensive model of communication behaviour during M&D team meetings to date. Comparisons between this and previous research is presented in Chapter 8.

## Interaction norms of construction progress meetings

Research objective 1 was to determine whether *the management and design meeting has a characteristic model of interaction, being different from the interaction profiles found in other contexts*. To determine whether the interaction behaviour of the construction professionals is different or similar to meetings in other contexts, the aggregated data is compared with previously published research results.

There are some trends that occurred repeatedly. Some interaction categories were consistent across all of the M&D meetings and others were prone to considerable variation between meetings. The most consistent observations are those that determine the characteristic patterns of interaction. Observations that exhibit differences are investigated in order to determine whether the difference observed resemble patterns of behaviour found in previous research. The first stage of this process was to examine and discuss the trends revealed by the descriptive statistics of the ten cases. Although there are trends that occur across all M&D team meetings, which are discussed later, it is important to note that most of the meetings within individual case studies exhibited trends that were specific to the particular project. For example, the meeting profiles in Case studies 9 and 10 exhibit little difference in the interaction behaviour within each case. However, other projects experience much greater levels of variation between meetings, particularly in the task dimensions. Figure 7.1 and Table 7.1 provides data that supports this discussion.

Three meetings were selected from each project for consistency. Chapter 5 provides evidence to suggest that the samples were representative of the cases

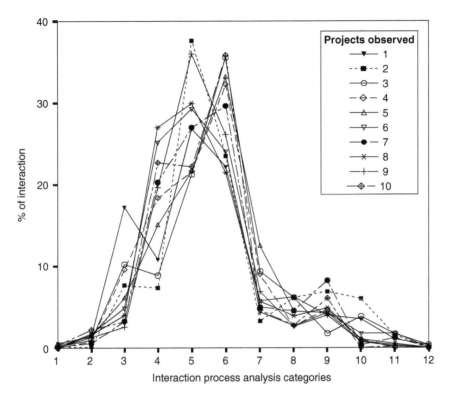

*Figure 7.1* Interaction profiles of all ten case studies

surveyed. An additional check of the profile produced from project data based on three meetings and the project data based on three and four meetings found no difference between the two data sets (Appendix 4), suggesting that the selected meetings were representative of the larger sample. The following line graph (Figure 7.2) provides a summary of group interaction observed during the M&D team meetings.

Figure 7.2 illustrates the general trend of group interaction found in the M&D team meetings. The norms operating within the meeting result in the groups giving task-based interaction being consistently higher than requesting task-based interaction. The positive socio-emotional interaction is also consistently higher than the negative socio-emotional communication. The degree of consistency varies depending on the category. The most consistent to least consistent categories are shown in Table 7.2. The more extreme socio-emotional categories (shows tension, shows tension release, shows antagonism, shows solidarity) were a rare occurrence in the M&D meetings, and the results in Table 7.2 show that this was consistent across most observations. So although many have described the construction sector as volatile and antagonistic, this was not evident in the data. Observations of extreme emotion were relatively consistent across the meetings.

Table 7.1 Results of all ten case studies and interaction categories

| Interaction categories | Project 1 | | | Project 2 | | | Project 3 | | | Project 4 | | | Project 5 | | |
|---|---|---|---|---|---|---|---|---|---|---|---|---|---|---|---|
| | No. | % | Adj. | No. | % | Adj. | No. | % | Adj. | No. | % | Adj. | No. | % | Adj. |
| 1: Shows solidarity | 6 | 1 | 4.1 | 0 | 0 | −1 | 2 | 0 | 0.2 | 0 | 0 | −1.5 | 4 | 0 | 1.2 |
| 2: Shows tension release | 16 | 1 | 0.4 | 1 | 0 | −3.0 | 14 | 1 | −1.1 | 13 | 1 | −1.8 | 30 | 2 | 1.1 |
| 3: Agrees | 198 | 17 | 15.4 | 61 | 8 | 1.3 | 146 | 10 | 6 | 157 | 10 | 5.4 | 118 | 6 | −0.8 |
| 4: Gives suggestion | 125 | 11 | −6.8 | 59 | 7 | −8.3 | 126 | 9 | −9.8 | 300 | 18 | 0 | 291 | 15 | −4.1 |
| 5: Gives opinion | 308 | 27 | −0.1 | 301 | 38 | 7.0 | 303 | 21 | −5.1 | 353 | 22 | −5.1 | 419 | 22 | −5.6 |
| 6: Gives information | 255 | 22 | −4.9 | 188 | 24 | −3.2 | 507 | 36 | 6.1 | 527 | 32 | 3.5 | 642 | 33 | 4.8 |
| 7: Asks for information | 66 | 6 | −1.6 | 26 | 3 | −4.2 | 133 | 9 | 3.8 | 147 | 9 | 3.5 | 242 | 13 | 10.4 |
| 8: Asks for opinion | 71 | 6 | 3.8 | 49 | 6 | 3.1 | 88 | 6 | 4.3 | 44 | 3 | −2.9 | 87 | 5 | 1.1 |
| 9: Asks for suggestion | 46 | 4 | −1.3 | 55 | 7 | 2.8 | 25 | 2 | −5.7 | 76 | 5 | −0.3 | 83 | 4 | −1.2 |
| 10: Disagrees | 40 | 4 | 5.2 | 48 | 6 | 10.1 | 55 | 4 | 7.1 | 16 | 1 | −2.1 | 15 | 1 | −3.1 |
| 11: Shows tension | 12 | 1 | 0.9 | 12 | 2 | 2.2 | 24 | 2 | 3.9 | 0 | 0 | −3.9 | 6 | 0 | −2.6 |
| 12: Shows antagonism | 4 | 0 | 4.5 | 0 | 0 | −0.7 | 4 | 0 | 3.9 | 0 | 0 | −1.0 | 0 | 0 | −1.1 |
| | 1147 | | | 800 | | | 1427 | | | 1633 | | | 1937 | | |

| Interaction categories | Project 6 | | | Project 7 | | | Project 8 | | | Project 9 | | | Project 10 | | | Total | |
|---|---|---|---|---|---|---|---|---|---|---|---|---|---|---|---|---|---|
| | No. | % | Adj. | No. | % | Adj. | No. | % | Adj. | No. | % | Adj. | No. | % | Adj. | No. | Adj. |
| 1: Shows solidarity | 0 | 0 | -1.7 | 1 | 0 | -0.7 | 0 | 0 | -1.4 | 1 | 0 | -0.7 | 4 | 0 | 2.2 | 18 | 0 |
| 2: Shows tension release | 37 | 2 | 2.2 | 9 | 1 | -2.9 | 25 | 2 | 1.2 | 21 | 1 | 0.1 | 27 | 2 | 3.1 | 193 | 1 |
| 3: Agrees | 101 | 5 | -3.3 | 55 | 3 | -5.6 | 62 | 4 | -4.2 | 41 | 3 | -6.8 | 39 | 3 | -4.8 | 978 | 7 |
| 4: Gives suggestion | 525 | 25 | 8.6 | 339 | 20 | 2.1 | 421 | 27 | 9.3 | 316 | 20 | 1.3 | 274 | 23 | 4.0 | 2776 | 18 |
| 5: Gives opinion | 611 | 29 | 2.6 | 451 | 27 | 0.1 | 466 | 30 | 2.8 | 580 | 36 | 8.7 | 268 | 22 | -3.9 | 4060 | 27 |
| 6: Gives information | 501 | 24 | -4.9 | 495 | 30 | 1 | 335 | 22 | -6.5 | 420 | 26 | -2.3 | 433 | 36 | 5.8 | 4303 | 29 |
| 7: Asks for information | 91 | 4 | -4.9 | 83 | 5 | -3.3 | 88 | 6 | -2.1 | 110 | 7 | -0.1 | 56 | 5 | -3.3 | 1042 | 7 |
| 8: Asks for opinion | 57 | 3 | -3.3 | 76 | 5 | 1.1 | 61 | 4 | -0.3 | 43 | 3 | -2.9 | 33 | 3 | -2.4 | 609 | 4 |
| 9: Asks for suggestion | 90 | 4 | -1.1 | 137 | 8 | 6.9 | 76 | 5 | 0.1 | 64 | 4 | -1.6 | 73 | 6 | 2.1 | 725 | 5 |
| 10: Disagrees | 37 | 2 | 0.6 | 5 | 0 | -4.5 | 16 | 1 | -1.9 | 11 | 1 | -3.1 | 0 | 0 | -4.6 | 243 | 2 |
| 11: Shows tension | 36 | 2 | 5 | 20 | 1 | 1.9 | 8 | 1 | -1.4 | 2 | 0 | -3.2 | 2 | 0 | -2.6 | 122 | 1 |
| 12: Shows antagonism | 0 | 0 | -1.1 | 0 | 0 | -1.0 | 0 | 0 | -1.0 | 0 | 0 | -1.0 | 0 | 0 | -0.8 | 8 | 0 |
| | 2086 | | | 1671 | | | 1558 | | | 1609 | | | 1209 | | | 15077 | |

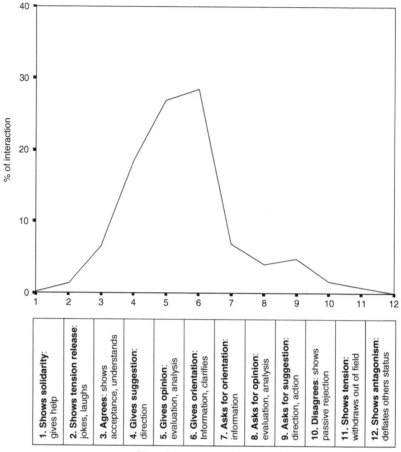

*Figure 7.2* Single interaction profile representing interaction data from 30 meetings, 3 meetings selected from each of the case studies

The level of agreement and disagreement, which is also described as a socio-emotional behaviour, is not consistent across meetings and experienced considerable variation. Categories that vary could be associated with unsuccessful and successful projects and are investigated further in Chapter 8.

### Comparison of interaction observed on each project

Project 2 had a larger variation between meetings than the other cases. A reason for the variation in meeting 2.3 was that one of the profiles represented the final M&D team meeting and the meeting exhibited a pattern that was not typical of other meetings in this project. Meeting 1.4 (Project 1) was also the final

*Table 7.2* Consistency of categories

| Interaction category<br>Most consistent to<br>least consistent | Adjusted residual (difference<br>between expected and observed<br>counts) |
|---|---|
| 2: shows tension release | 3.1 |
| 1: showing solidarity | 4.1 |
| 8: asking for opinion | 4.3 |
| 12: shows antagonism | 4.5 |
| 11: shows tension | 5.0 |
| 6: gives orientation | −6.5 |
| 9: asks for suggestion | 6.9 |
| 5: gives opinion | 8.7 |
| 4: gives suggestion | −9.8 |
| 10: disagrees | 10.1 |
| 7: asks for orientation | 10.4 |
| 3: agrees | 15.4 |

meeting of the project. Graphs 1.4 and 2.3 (Figures 6.1 and 6.2) have similarities between them that are qualitatively different from the other meetings observed in this research. The results indicate that final M&D team meetings have particular interaction tendencies, but only two of the meetings were the final M&D meeting, so there is insufficient data to provide strong support for this proposition. Issues surrounding the interaction associated with project closure would be worthy of further research.

A number of interaction patterns were reproduced in many of the projects, although some were peculiar to an individual project. In all projects, except Project 2, the meeting interaction patterns within each project are similar. This suggests that the groups associated with each project had established their own group interaction norms. Differences between each of the case studies were often confined to a limited range, which was substantially reduced if the upper and lower extreme values were removed. The amount of deviation from the mean varies depending on the individual IPA category. A few of the IPA categories exhibit little deviation from the mean, throughout the projects and meetings. It is now possible to examine in detail where the most consistent patterns emerged and where the greatest departures from the mean were found.

### Category use and consistency across M&D team meetings

Category 1 results account for 1 per cent, or less, of the total interaction within each case. So it is not common to see showing solidarity, raising others status, giving praise or reward type communication acts during the meetings. However, the observations of this category were associated with projects that had successful outcomes (within budget, within time and no conflict). In all of the projects that were deemed successful, this extreme behaviour of positive socio-emotional support was observed. In contrast, the projects that were less successful did not experience such behaviour, or only experienced it on a single occasion. The data, although limited, does provide some indication that support systems on successful projects are more developed.

The interaction observed in category 2 (showing tension release) accounts for just over 1 per cent of the total interaction. When this category was observed it occurred at the beginning or the end of meetings, helping to ease any tension in the group. It also occurred at the end of discussions when tension was noticeable or the amount of disagreement was high. Interaction associated with category 2 is consistently low; it was uncommon for high levels of tension release to occur in the meetings. Despite being low, interaction in this category occurred more than it did in category 1.

Category 3 (agreeing) occurred more than the other two positive socio-emotional categories (showing solidarity and showing tension release). Interaction in this category was the least consistent of all categories. The most extreme observation was that found in Project 1, where agreement accounted for 17 per cent of the interaction. Generally the higher levels of agreement are associated with successful projects. Regardless of this, the amount of agreeing is low, generally representing 10 per cent, or less, of the interaction observed. Agreement could occur for a number of reasons. Agreement was often used to reassure parties that the information provided or the progress reported was that which was expected, such support encouraged parties to continue with their dialogue and reinforced the value of their contribution. Agreement occurred the most where suggestions were given in response to a difficulty or action point. The level of agreement and the strength of sentiment surrounding the discussion recognised the importance of the progress that the members in the meeting had made on a particular issue.

The task-based information associated with giving suggestions (category 4), giving opinions (category 3) and giving information (category 6) had some of the highest levels of variation across the interaction categories. These categories were also those most used by the M&D groups. Giving suggestions was the lowest of the 'giving task-based' interaction categories. With the exception of Project 2 and Project 3, which are considerably lower than the others, and Project 8, which is high, the remaining seven cases have a variance of less than 8 per cent.

Information type communication acts were used to provide background data, contextualise and present facts related to problems. Opinion and analysis acts were used to explore issues and offer different perspectives. Opinion based acts were also used to sensitively offer suggestions, from an individual perspective, without implying that the suggestion was the most appropriate idea. Opinions that were asserted with authority or implied direction, suggesting that this was the way things were to be done, were categorised as suggestions. Interaction categorised as giving suggestions was often used to direct discussions and action. Such acts were usually autonomous statements, with the person making the suggestion attempting to adopt a leading role in the group discussion, or attempting to bring discussions to a close. The multidisciplinary nature of the group meant that many different perspectives could be offered. Supporting information and explanation was needed so that others could understand the discussion. Prominent use of suggestions was usually limited to one or two members in the group. The use of suggestions had the effect of changing the nature of interaction from passive discourse, where information was merely exchanged to aid understanding, to assertive and directive behaviour. A hierarchy of task-based communication exists. Some communication acts such as exchanging background information are passive, while suggestions and direction are much more assertive communication acts. As interaction changes from passive to assertive, those engaged become more attentive and focused on the interaction. The length of assertive discourse varied. Assertive behaviour was rarely prolonged, such episodes being broken up by exchanges of background information and agreement.

Categories 5 (giving opinions) and 6 (giving suggestions) produced similar levels of interaction. Although the total interaction observed in category 6 is higher than category 5, in half of the individual case studies category 5 is higher than category 6. The interaction observed in category 6 and 5 in most cases falls between 25 and 35 per cent. High levels of giving information and opinions were consistent across the meetings.

Requesting information (category 7) experienced some variation, as did asking for suggestions (category 9), but to a lesser extent. Asking for opinions (category 8) experienced the lowest variation of any of the task-based categories. The level of interaction observed in the categories that requested a response was much lower than the other giving task-based categories. Participants were five times more likely to use giving communication acts than to ask for information, opinions or suggestions.

Asking for information had the largest variation of the requesting type acts; however, disregarding the extreme value in Project 5, the variation was low (4 per cent). Asking for information was the most used of the 'requesting' categories. Asking for opinions was consistent across all cases, although it only represented 4 per cent of the total interaction. Asking for suggestion experienced greater variation between case studies than asking for opinions; the interaction for this category accounts for 5 per cent of the total interaction observed in all case studies.

The use of requesting interaction behaviour varied. Although used quite sparingly, it was used to ask for further information and opinions; however, it was rarely used to ask for explanation because a member was unable to understand. Indeed, explicitly asking for help was not observed during any of the meetings. In Project 9, a female participant implied that help was required, and sought a suggestion from another group member. In this particular case the help-seeking behaviour was used to raise the status of the person from whom the information was requested, and this person was a rather reluctant contributor.

Requesting suggestions was occasionally used to insist that another group member offered solutions. Such acts were used to bring an end to exchanges of information and opinions and to encourage others to commit to a direction.

The total amount of negative socio-emotional interaction (categories 10–12) was low, being much lower than positive socio-emotional interaction. Disagreements (category 10) manifested the greatest variation of the negative socio-emotional categories. The highest level observed within a case was 6 per cent

(Project 2) and the lowest was less than 1 per cent in Project 10. The levels of disagreements were consistently low; arguments and challenges were not common.

The more extreme acts of negative socio-emotional interaction were very low. The highest level of showing tension (category 11) was 2 per cent (Case studies 3 and 6), and the total interaction recorded was only 1 per cent. All except one of the cases experience levels of tension, although the levels of tension within the M&D team meetings are consistently low. Observations of the extreme socio-emotional communication categories (showing antagonism) during M&D meetings were rare and only occurred in two projects.

The amount of negative emotion in the meetings was low, and the expression of tension was much lower. Those who dominated interaction were also more inclined to show when they disagreed and expressed tension; less proactive interactors hardly disagreed with others. In most cases disagreement and the expression of tension were directed at a specific problem or directly at another member's argument or proposal; however, in one of the case studies many of the negative socio-emotional acts were used to express frustration. The lack of direction of the frustration did not result in other members attempting to defend a proposal or position. During the meeting, the expression of frustration did not seem to result in others offering any corrective action or justifying their position. However, when negative interaction was directed at a problem or another person, interaction quickly emerged to defending initial proposals or positions. Such exchanges often resulted in different courses of action from that initially proposed.

Although variation in some of the individual categories is quite high, there is little variation between the aggregated results that form the task (categories 4, 5, 6, 7, 8 and 9) and socio-emotional (categories 1, 2, 3, 10, 11 and 12) groups. The ranking of the task and socio-emotional groups does not change in any of the ten cases observed.

All ten case studies, regardless of project success, have higher levels of positive socio-emotional communication (categories 1, 2 and 3) than negative socio-emotional communication (categories 10, 11 and 12). In Case study 2, the difference between the positive and the negative socio-emotional communication is small but slightly higher positive direction. If all the cases are aggregated under these levels of interaction, the positive and the negative socio-emotional communication behaviour accounts for 8 and 3 per cent respectively.

Another aggregated aspect considered is the relationship between requesting (categories 7, 8 and 9) and giving task-based interaction (categories 4, 5 and 6). In all of the case studies the amount of task-based interaction given exceeds the amount of task-based interaction requested by a minimum of three times. Taking the aggregated data for all cases, giving and receiving task-based interaction accounts for 74 and 16 per cent respectively. The two relationships identified, positive exceeding negative socio-emotional communication, and giving exceeding requesting task-based interaction, are consistent throughout all of the case studies.

Figure 7.2 shows the profiles for all ten cases. Less deviation occurs in categories 1, 2, 6, 8, 9, 11 and 12 (see Table 7.1) than other categories. Most of these categories 1, 2, 11 and 12, represent the more extreme expressions of emotional behaviour. Categories 1 (showing solidarity), 2 (showing tension release), 11 (showing tension) and 12 (showing antagonism) do not often occur, and this behaviour is consistent throughout meetings.

## Comparison of findings with previous studies

Previous studies of group communication have tended to discuss the relationship between positive and negative socio-emotional communication. For positive socio-emotional interaction, the values of categories 1, 2 and 3 are combined and for negative socio-emotional interaction, the values for categories 10, 11 and 12 are aggregated.

Our results suggest that it is not normal to express high levels of extreme emotional behaviour during M&D meetings. Bales' studies of adult academic groups (Figure 7.3) found that the low levels of socio-emotional behaviour were

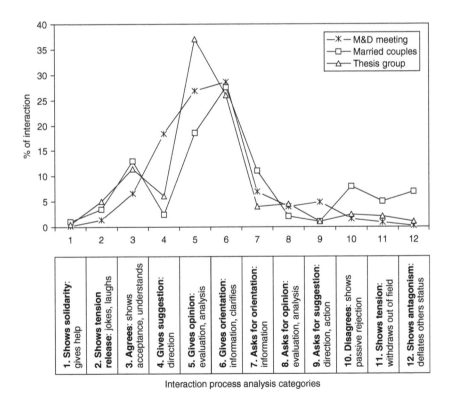

*Figure 7.3* M&D profile and Bales' (1950: 25) observations of married couples and thesis groups

far removed from the types of behaviour that would be found in infant groups. Communication in adult groups is very structured, controlled and restrained (Bales 1950, 1970). In the M&D meeting the levels of emotional interaction were low and lower than that reported by Bales. We suspect that the commercial environment reduces emotional communication to a level lower than that found in counselling groups, social groups and laboratory experiments. Gameson (1992) found lower levels of socio-emotional communication than those found here. Bellamy *et al.* (2005) also found low levels of socio-emotional interaction in both co-located and virtual environments (Figure 7.4); this too was lower than Bales' earlier research. The level of socio-emotional communication in the design teams is generally comparable to that of a mixed group of M&D professionals; although the task-based interaction is different.

As groups develop so their behaviour changes. Groups initially use task-based communication (Bales 1950) tentatively gathering information on other's

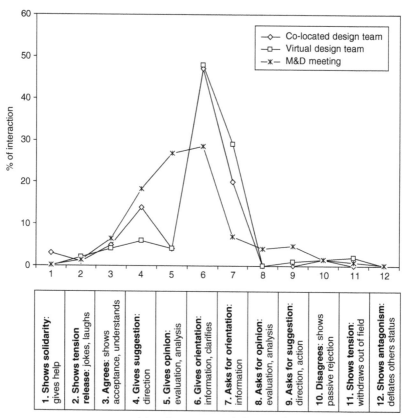

*Figure 7.4* Interaction in M&D meetings and co-located and virtual design team interaction (Bellamy *et al.* 2005: 359)

personal beliefs and attitudes before using and developing socio-emotional interaction. Once groups reach the later stages of group development the amount of socio-emotional communication increases (Heinicke and Bales 1953; Bales 1970). Gameson's (1992) research supports this view: in a one-off meeting between clients and construction professionals the majority of interaction observed by Gameson was task-based (Figure 7.5). Observations of showing solidarity, showing tension release, showing tension and antagonism were negligible; agreeing and disagreeing during this 'one-off' meeting were also low. The M&D group does not experience heightened levels of socio-emotional communication because it does not reach the final levels of group development.

## Differences in interaction: To what extent is the M&D team peculiar

To determine whether the interaction behaviour is peculiar to the M&D meeting, other studies of groups are compared and any differences noted. The data for the M&D meeting is based on a total of 30 meetings (Figure 7.2). The following graphs include the M&D meeting line graph superimposed on to Bales' (1950) profiles (Figure 7.3), Bales' (1970) later review of 21 studies

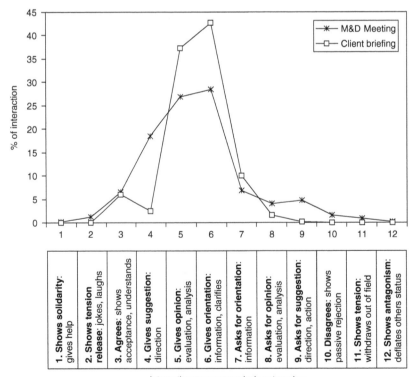

*Figure 7.5* M&D profile and Gameson's (1992: 312–317) observations of client briefing

of adult groups, which provides a range of adult norms (Figure 7.6), Cline's (1994) study of disagreement and agreement (Figure 7.7), Landsberger's (1955) records of management disputes during 12 mediation meetings (Figures 7.8) and Bell's (2001) observations of multidiscipline child protection teams (Figure 7.9). Gameson's (1992) and Bellamy *et al.*'s (2005) results, which are specific to the construction industry, are also considered (Figures 7.5 and 7.4).

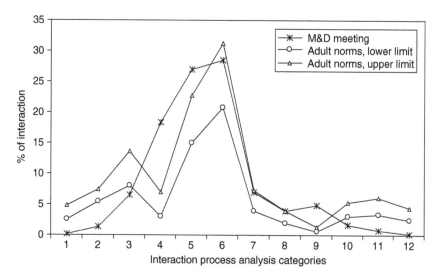

*Figure 7.6* M&D profile and Bales' (1970: 473) interaction norms of adult groups

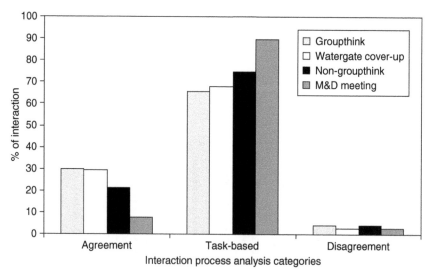

*Figure 7.7* M&D and Cline's (1994: 214) observations interaction that results in 'groupthink'

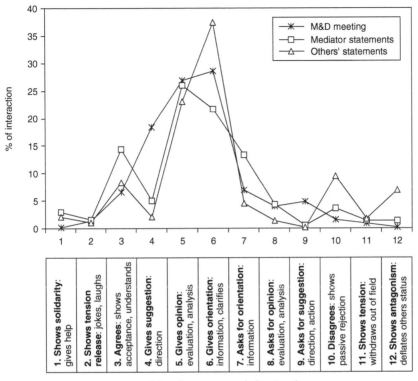

*Figure 7.8* M&D meeting profile and Landsberger's (1955: 568) observations of workers' representatives and management dispute during mediation

While the interaction profiles of children may have little to do with this work, they present an example of what is described as 'unorganised communication' (Bales 1950). The socio-emotional behaviour of children is relatively uninhibited and uncontrolled when using task-based interaction, and the suggestions made by the children are presented almost entirely unsupported by accompanying analysis, inference or persuasion (Bales 1950). Socha and Socha (1994) investigated changes in children's behaviour and noted that as discussions continued there was an increase in task and reduction in socioemotional interaction, moving towards the behaviour of adults. In adult groups, information and suggestions are normally supported by explanation. It is important to note that the M&D meeting interaction is much more restricted and bears little resemblance to children's behaviour, being more typical of adult groups, although differences to previous adult studies are noted. Even in the M&D meeting observed in Project 7, where a level frustration was expressed without identifying the exact reason for the frustration, the emotional

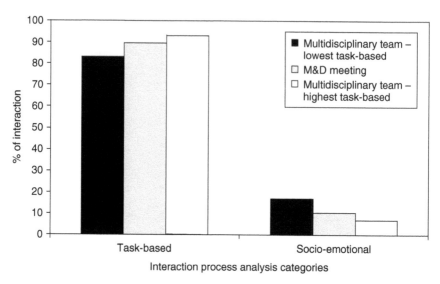

*Figure 7.9* Interaction in M&D meetings and multidisciplinary child protection team meetings (Bell 2001: 74)

interaction patterns produced do not resemble the type of behaviour used by children.

As already discussed, the levels of social-emotional behaviour in M&D meetings are very low. This type of behaviour is more comparable to the academic discussion group studied by Bales (1950: 26) (Figure 7.3), where the amount of socio-emotional behaviour was described as a bare minimum, 'being as far removed from pre-school child behaviour as one could imagine'. While categories 1 (showing solidarity) and 2 (showing tension release) are low in the M&D profiles, Bales (1950) stated that in adult discussion groups we do not ordinarily find the rate of joking and laughing to be higher than the rate of agreement.

Although the positive socio-emotional communication in the thesis group is described as a minimum, there is even less positive socio-emotional communication in the M&D team meeting results. The results show that the construction participants are reluctant to use socio-emotional communication in this M&D team meeting. It is noted that groups tend to use more neutral socio-emotional interaction during early stages of interaction, with positive and negative social interaction being increasingly used in the later phases of group development (Bales 1950, 1953). The low level of socio-emotional interaction observed in the M&D team meeting is consistent with the results from Bellamy *et al.*'s (2005) research.

Wallace (1987) (mindful that he adapted the Bales IPA method) found that when the group changed (members entered or left the group), the group regressed back to early stages of the group interaction development process. During the M&D team meetings the majority of members were consistent, but new members would occasionally join the group. This resulted in temporary changes to interaction, as people familiarised themselves with other actors and the topic being discussed. Considering the complexity of the building and associated discussions, such familiarisation periods were short and the behaviour typical in previous meetings soon resumed. The number of participants attending each meeting varied (Table 6.1), although 7–10 members generally attended each meeting. Bales' (1950) report of different sized chess groups showed that the number of participants undertaking a specific activity does not necessarily have a great effect on group interaction patterns (discussed in Chapter 3). Differences attributable to the variation in the group size were not found in the M&D team meetings.

Gameson's results (Figure 7.5) provide data on the early stages of group behaviour, but there is no recorded evidence of the extreme socio-emotional categories 1, 2, 11 and 12. Results were generated from transcripts of audio tape recordings, which may limit such observation, the only socio-emotional categories observed were agreeing and disagreeing. The levels of agreeing were comparable to those found in the M&D meeting; however, Gameson's observations of disagreeing were considerably lower than our findings (Figure 7.5). Professional relationships in this and Gameson's study have not developed sufficiently enough to allow higher levels of socio-emotional interaction to emerge.

The findings suggest that disagreements during initial meetings are rarely observed. The level of disagreement is comparable to that found in Cline's (1994) (Figure 7.7) study of differences between the interaction of groups that fall into the trap of groupthink and those groups that avoid groupthink. Cline's study suggests that group members who agree with each other without challenging and evaluating proposals can suffer from groupthink. The important point here is that the level of agreement, although less than found in Cline's study, is closer to groups that do not suffer from groupthink.

Comparison between Bales (1970), Landsberger's (1955) and the M&D team results (Figure 7.8) is illuminating. In the M&D meetings the group did not engage in the same level of disagreement as that previously reported by Bales, nor did it reach that experienced in Landsberger's study of disputes. Landsberger found that those unable to resolve their differences on their own and who subsequently used a third party, such as a mediator, expressed a higher level of disagreement, tension and antagonism during mediation than found in M&D meetings. Disagreement is also lower than that reported in Cline's study of groupthink – failing to disagree is an indicator of groupthink. Rather than being classed as adversarial, the low level of disagreement could be detrimental to the groups' performance. If members are accommodating the views of others without challenging them, when they secretly disagree or have concerns with the proposal, then they fall victim to groupthink. If groupthink occurs, problems often manifest and become of greater consequence as time unfolds. Although the level of disagreement is low, so is the level of agreement. The level of agreement is closer to the groups studied by Cline that do not show traits of groupthink. The issue of disagreement and groupthink requires further investigation.

The use of negative emotional expression had a profound effect on the groups' behaviour. When members of the group disagreed or expressed tension the mood of the discussion quickly changed; the members became much more focused on the conversation. Those engaged in the disagreement held the floor while others remained silent. Those with an interest in the outcome of the situation would interject to reinforce the importance of the issues discussed or attempt to control emotion by suggestions that helped resolve matters. Although a rare occurrence, the effect of negative emotion (if focused) could be used to ensure matters were given desired importance and resulted in action or agreement. Negative and positive emotion had a considerable effect on the group behaviour. The use of emotional communication acts and signals changed the group dynamic. Some members used socio-emotional interaction to influence group behaviour and reaction.

The level of socio-emotional interaction observed by Cline (1994), Gameson (1992), Bell (2001) and Bellamy *et al.* (2005) are notably lower to that found by Bales (1970). Interaction between professionals (in business rather than social situations) uses less socio-emotional interaction. The formal interaction between construction professionals shows greater emotional restraint when compared to group discussions and meetings in other contexts. Even in the context of a labour–management dispute found in Landsberger's (1955) results, the level of socio-emotional interaction is lower than Bale's (1970) group norms would indicate. Early research by Bales, which did look at a multitude of meetings and group discussions, tended to look at the behaviour of school groups, college students, family discussions, groups in roll play situation and other social situations. More recent studies that have looked into group behaviour in the business context show considerable differences in the interaction style.

In the business context, there is greater use of the giving task-based information and far less use of the socio-emotional categories. In the M&D meeting, when socio-emotional communication acts were used they often resulted in reciprocal responses of emotional behaviour. When positive emotion was used most of the participants responded with reciprocal positive emotional expression. The occurrence of positive emotion stimulated, influenced or encouraged a positive emotional response from other participants. The heightened positive expression continued when all group members participated or was restrained and normality restored where key members did not participate with the same level of positive emotion. It should be noted that positive expression from prominent and influential members was often more restrained. The restrained behaviour of a few members has the effect of quickly restoring the normal task-based interaction. Allowing the emotional expression, but then curtailing it ensured that the meeting was emotionally expressive yet sufficiently reserved to ensure that it dealt with business matters with relevant rigor. Short emotional episodes were often used to quickly disperse tension or ease the group discussion.

## Discussion of team interaction characteristics

Both the results presented here and those of previous research suggest that the low level of socio-emotional interaction in the M&D meeting is resulting from the temporary nature of the team, its instability (with people entering and leaving the group) and the short period of time that the group has existed. Thus, the group has not progressed to the later stages of group development achieved by other groups, or else the group restrains itself from using socio-emotional expression. The level of emotional restraint and control that key members of the group showed would suggest that they use the small levels of emotion to influence group behaviour.

Gameson (1992) found that the differences in the use of task-based categories were related to experience and profession. Both experienced and inexperienced clients, when meeting with construction professionals, primarily used category 6 (giving information); however, the inexperienced clients tended to request more information (category 7) and opinions (category 8). In this research the total interaction observed shows higher levels of information giving (category 6) and marginally lower levels of opinion giving (category 5); however, in half of the cases the level of opinions were higher than information (Table 7.1).

The pooled interaction of five married couples (Figure 7.3) has much higher levels of emotional communication; however, the task-related categories giving information (category 4) opinions (category 5) and suggestions (category 6) while not as high, follows the same order as the M&D group. Bales (1950) inferred that the high level of antagonism (category 12) observed between married couples is only possible due to the strength of the relationship, knowing that such emotional outbursts would not damage the relationship. Therefore, it is not surprising that the extreme negative emotional outbursts are uncommon in the M&D team meetings, and other commercial or 'working' environments.

The levels of giving information (category 6) of the M&D profile are higher than giving opinion, analysis etc. (category 5). In the studies described (Bales 1950: 19–29), this type of pattern, with giving information being higher than opinions, is presented in only one situation: interaction amongst the groups of five married couples. It was suggested that this pattern could arise from a type of communication efficiency. The close relationship allows them to simply state relatively incomplete information that would otherwise need to be explained. In the M&D meeting, construction professionals use construction terms that have implied meaning, without the need to fully explain them. Construction professionals are able to do this due to a degree of commonality in education, training and experience. This may help to explain the high level of information given. However, as most projects are temporary in nature and participants belong to different organisations, it is unlikely that the professionals have such a close relationship that all statements carry additional information that is 'fully' understood. In the earlier studies of professionals the level of information exceeds the level of opinions. The high level of giving information, opinions and explanation is considered to be necessary to enable parties to express issues sufficiently so that other members understand; however, it is questionable why levels of explanation are not greater than information.

Gameson (1992) also observed that the construction professionals used high levels of category 6, when inexperienced clients and construction professionals met for the first time. The information provided was supported with opinion, explanation and analysis. When the construction professionals met with experienced clients, the amount of category 5 (opinions and explanation) was higher than category 6 (information). Clients who understood less about the construction process received less information than the more knowledgeable and experienced clients. With experienced clients, the construction professionals had to justify their information and suggestions more than with inexperienced clients. These results are interesting because the level of asking for explanation and opinions (category 8) by inexperienced and experienced clients was similar. The only difference in the level of asking question was that experienced clients asked for more information, but received more opinions and explanation than information.

Bales' (1970) profiles shown in Figure 7.6, while bearing some similarity to the M&D profile, are higher in the amount of positive and negative socio-emotional interaction. The task-related interaction of these profiles differs most in categories 4 and 6, giving information, opinions and suggestions.

## Summary

The positive socio-emotional categories increased from category 1 to 3. The most extreme expression of positive socio-emotional interaction, category 1, has the lowest percentage. This observation occurs in all M&D meetings observed. Such behaviour is also typical of all of the adult groups observed by Bales (1950, 1970). In comparison with other studies the amount of positive socio-emotional interaction observed is considerably less (apart from Gameson's research and other studies that involve professionals). The amount of negative socio-emotional communication is also lower. A combination of the members' temporal relationships and the commercial context in which they take place appears to restrain interaction, thus restricting socio-emotional development, as achieved in more stable groups. The high levels of conflict reported in construction literature were not evident in the data; indeed, if the levels of conflict reduced much further, the groups may be subject to the occurrence of groupthink. As Loosemore *et al.* (2000) suggest, there may be too much emphasis on conflict avoidance and that the current indiscriminate policy towards conflict reduction could result in positive process losses. When actors blindly agree without attempting to disagree or challenge others, while secretly being mindful of potential problems, then groupthink is manifest.

A number of peculiarities were found in the task-based interaction categories. Giving suggestions, giving information and requesting suggestions have higher interaction levels than any of the Bales' studies. Category 6 was lower than Gameson's study. Although the requests for task-based interaction are not considerably different from the previous studies, all of the three giving task-based interaction categories are high and the amount of giving compared with requesting task-based interaction is greater than all of Bales (1950) studies. The high level of task-based interaction resulted from the low level of socio-emotional interaction, due to the groups' restricted development. This is supported by Gameson's (1992) results, which have even higher amounts of task-based interaction than the M&D meeting.

# 8 Successful and unsuccessful project outcomes

The second objective of our research was to determine whether any differences existed between the group interaction associated with successful projects and that associated with unsuccessful projects. Hypothesis 1 states that *the interaction patterns of group meetings associated with successful project outcomes are significantly different from the interaction patterns of group meetings that are associated with unsuccessful project outcomes.*

Feedback from contractors on their projects' performance indicated that seven of the projects were completed within budget and did not experience conflict with the client; five were completed within the scheduled duration and only one case resulted in conflict with the architect and main contractor. All of the projects that did not complete within budget also experienced conflict between the client and the main contractor. The graphs and tables that present the data for the budget are the same as those that would be generated from the data on conflict with the client. For ease of discussion, only one table will be produced for these two outcomes (graphs for conflict between client and contractor and graphs that show projects that did not complete within budget). From the interaction data generated, the group interaction associated with successful projects (those with positive outcomes) can be compared with unsuccessful projects.

## Different levels of interaction analysis

### Socio-emotional and task-based interaction

Interaction data gathered using the Bales IPA can be investigated from a number of different perspectives. In this research the first level of analysis was the emotional and task-based components of communication and their relationship with project outcomes. Two categories were created by aggregation of data from the six emotional communication acts (shows solidarity, shows tension release, agrees, disagrees, shows tension and shows antagonism) and the six task-based acts (gives suggestions, gives opinions, gives information, asks for orientation, asks for opinions and asks for suggestions).

*Positive and negative socio-emotional, and giving and requesting*
*task-based categories*

The second level of analysis looked at the emotional- and task-based components in more detail and examined whether the four interaction categories associated with the positive and negative socio-emotional interaction, and giving and requesting task-based interaction were significantly different when examined against the project outcomes. To examine the socio-emotional and task-based categories, individual IPA categories were grouped together. Positive socio-emotional categories comprise of the aggregated data from categories 1 (shows solidarity), 2 (shows tension release) and 3 (agrees). Negative socio-emotional categories were compiled from the data in categories 10 (disagrees), 11 (shows tension) and 12 (shows antagonism). The task-based categories were also split into two groups: giving and requesting task-based interaction. Giving task-based interaction included categories 6 (gives information), 7 (gives opinions) and 8 (gives suggestions). Requesting task-based interaction included categories 7 (asks for orientation), 8 (asks for opinion) and 9 (asks for suggestion).

*Individual IPA categories*

The third level of group interaction analysis is the most detailed and has been used to determine whether the 12 categories of interaction are different in the groups associated with successful and unsuccessful project outcomes. Where significant differences were found, the line graphs and tables are used to determine the direction and scale of difference between the data. In addition to examining all IPA categories together, within one data set, each category was also considered separately.

## Socio-emotional and task-based interaction

In this examination, emotional communication includes all acts that express emotion, and the task-based interaction was composed of all of the giving and requesting task-based communication acts. Tables 8.2.1–8.2.3 provide a comparison and statistical analysis of task-based and socio-emotional interaction in situations of successful and unsuccessful project outcomes. The Pearson Chi-square statistic $(\chi^2)$ was used to test the relationship between interaction levels and project outcomes. The findings were reported as probability values $(p)$, the degrees of freedom are also stated. The findings were used as evidence supporting or rejecting of the hypothesis. Here the hypotheses were based on the acceptance that the level of interaction was related to the project outcome (thus, a significant difference existed between the interaction associated with successful and unsuccessful projects). In this current work a significant difference was recorded as $p = < 0.05$, although, due to the level of multiple analysis carried out, the significance of this result was reduced. A more reliable indicator of significance is a $p$ value of $< 0.01$ and results of $< 0.001$ are considered to

*Table 8.2.1* Results: Projects within budget and over budget–Socio-emotional and task-based interaction. *(Conflict with client was present in all cases that were over budget.)*

| Interaction categories | Within budget | | Over budget | | Total interaction observed | |
|---|---|---|---|---|---|---|
| | No. | % | No. | % | No. | % |
| Task-based interaction | 8 612 | 89 | 4 903 | 91 | 13 515 | 90 |
| Socio-emotional interaction | 1 075 | 11 | 487 | 9 | 1 562 | 10 |
| Total interaction observed | 9 687 | | 5 390 | | 15 077 | |

Notes
$\chi^2 = 15.857$, $df = 1$, $\rho = <0.001$.

*Table 8.2.2* Results: Projects within and exceeding scheduled duration–Socio-emotional and task-based interaction

| Interaction categories | Within schedule | | Over schedule | | Total interaction observed | |
|---|---|---|---|---|---|---|
| | No. | % | No. | % | No. | % |
| Task-based interaction | 5 305 | 90 | 8 210 | 90 | 13 515 | 90 |
| Socio-emotional interaction | 597 | 10 | 965 | 11 | 1 562 | 10 |
| Total interaction observed | 5 902 | | 9 175 | | 15 077 | |

Notes
$\chi^2 = 0.627$, $df = 1$, $\rho = 0.429$.

*Table 8.2.3* Results: Projects that did and did not experience conflict with the architect–Socio-emotional and task-based interaction

| Interaction categories | No conflict | | Conflict | | Total interaction observed | |
|---|---|---|---|---|---|---|
| | No. | % | No. | % | No. | % |
| Task-based interaction | 11 640 | 90 | 1 875 | 90 | 13 515 | 90 |
| Socio-emotional interaction | 1 351 | 10 | 211 | 10 | 1 562 | 10 |
| Total interaction observed | 12 991 | | 2 086 | | 15 077 | |

Notes
$\chi^2 = 0.157, df = 1$, $\rho = 0.692$.

be highly significant. Results of this nature indicate that there is a significant difference between interactions.

In two of the three comparisons (Table 8.2.1, 8.2.2 and 8.2.3), emotional interaction associated with successful outcomes was greater than the emotional interaction of unsuccessful outcomes, thus the task-based communication was lower in situations of positive outcomes. However, in two of the situations, projects within and exceeding scheduled duration ($\chi^2 = 0.627$, $df = 1$, $\rho = 0.429$) and projects that did or did not experience conflict with the architect ($\chi^2 = 0.157$, $df = 1$, $\rho = 0.692$), the difference between emotional and task-based interaction was not significant. This shows that the general relationship between task-based and socio-emotional interaction in successful and unsuccessful projects is relatively consistent. If there is a difference, it occurs within the categories that make up task-based and socio-emotional interaction.

In the situation where the difference was significant ($\chi^2 = 15.857$, $df = 1$, $\rho = <0.001$), projects within and exceeding budget, the levels of emotional communication were higher in the cases associated with positive project outcomes (Table 8.2.1). Although the difference shown by the Chi-square test is greater than would be attributable to chance, the differences in the actual percentages of interaction are very small.

The comparisons of emotional and task-based interaction for successful and unsuccessful projects did not show a consistently strong trend. At this aggregated level of analysis, no useful inferences can be deduced from the results. The groups associated with positive and negative project outcomes used similar levels of emotional and task-based interaction regardless of whether they were successful or not.

## Positive and negative socio-emotional interaction

Early indications show that greater amounts of positive socio-emotional inter-action compared with negative socio-emotional interaction are desirable, having a stronger association with the more successful groups. The interaction data presented in Table 8.3.1 from the projects completed within budget exhibit slightly higher positive socio-emotional communication (categories 1: showing solidarity, 2: tension release and 3: agreeing) than the projects that were over budget, and these differences are significant. The projects that were within budget

*Table 8.3.1* Results of positive and negative socio-emotional interaction, and giving and requesting task-based interaction: Projects within budget and over budget

| Interaction categories | Within budget | | Over budget | | Total interaction observed | |
|---|---|---|---|---|---|---|
| | *No.* | *%* | *No.* | *%* | *No.* | *%* |
| Positive socio-emotional interaction | 816 | 8 | 373 | 7 | 1 189 | 8 |
| Giving task-based interaction | 7 037 | 73 | 4 102 | 76 | 11 139 | 74 |
| Requesting task-based interaction | 1 575 | 16 | 801 | 15 | 2 376 | 16 |
| Socio-emotional interaction | 259 | 3 | 114 | 2 | 373 | 3 |
| Total interaction observed | 9 687 | | 5 390 | | 15 077 | |

Notes
$\chi^2 = 24.202$, $df = 3$, $\rho = <0.001$.

used less 'giving task-based communication acts' than the projects that were over budget (73 per cent compared with 76 per cent). Projects within budget also used higher levels of requesting task-based communication acts than projects that were over budget. The results of the Chi-square test for projects within and over budget is much greater than would be attributable to chance ($\chi^2 = 24.202$, $df = 3$, $\rho = <0.001$).

Table 8.3.2 presents the interaction data for projects completed within the scheduled duration or not. This table presents similar results to those of the cost comparison. The projects that were completed within the scheduled duration used higher levels of positive socio-emotional communication than projects that failed to complete within the scheduled duration.

The amount of positive socio-emotional communication for projects completed within the scheduled duration, and those projects that exceeded the scheduled duration, accounted for 9 and 8 per cent of the interaction respectively, and the difference between these two figures was significant. The amount of negative socio-emotional interaction of the projects that exceeded the project duration in this analysis was 1 per cent higher than those that completed within time. The projects that were completed on time also have 7 per cent more positive than negative interaction, whereas projects that exceeded the scheduled duration only used 5 per cent more positive than negative interaction. These figures continue to support the argument for higher levels of positive rather than negative socio-emotional interaction, this being associated with successful project outcomes.

Consistent with the cost analysis (Table 8.3.1), the projects that were completed within the scheduled duration (Table 8.4.2) used less 'giving task communication acts' than the projects that were in excess of the scheduled duration (73 per cent compared with 74 per cent). Also consistent with the previous cost results, projects completed within the scheduled duration use a higher proportion of communication associated with requesting task-based acts than projects over budget (17 per cent compared with 15 per cent). All of the differences at this level were significant.

*Table 8.3.2* Results of positive and negative socio-emotional interaction, and giving and requesting task-based interaction: Projects within and over scheduled duration

| Interaction categories | Within schedule | | Over schedule | | Total interaction observed | |
|---|---|---|---|---|---|---|
| | *No.* | *%* | *No.* | *%* | *No.* | *%* |
| Positive socio-emotional interaction | 505 | 9 | 684 | 8 | 1 189 | 8 |
| Giving task-based interaction | 4 331 | 73 | 6 808 | 74 | 11 139 | 74 |
| Requesting task-based interaction | 974 | 17 | 1 402 | 15 | 2 376 | 16 |
| Socio-emotional interaction | 92 | 2 | 281 | 3 | 373 | 3 |
| Total interaction observed | 5 902 | | 9 175 | | 15 077 | |

Notes
$\chi^2 = 42.090$, $df = 3$, $\rho = <0.001$.

In the earlier table (Table 8.2.2), which restricted the comparison to emotional and task-based acts, no significant difference was found between projects that were completed within and exceeded the scheduled duration. However, in this situation when the task-based categories are divided into giving and receiving task-based interaction and positive and negative socio-emotional acts, the difference was highly significant ($\chi^2 = 42.090$, $df = 3$, $p = <0.001$). Thus, at a more detailed level of investigation into the task-based and socio-emotional acts, we find some useful difference.

Differences are emerging in the groups using task and socio-emotional communication acts. Less successful groups tend to devote a slightly higher proportion of their interaction to information giving and have lower levels of asking for task-based and positive socio-emotional communication acts. The successful groups use more positive emotional expression and request more from other members of the group.

### Conflict with the architect

Only one of the ten case studies resulted in conflict between the architect and the contractor's representative. Thus, the strength of any argument that is associated with projects that result in conflict with the architect is limited. However, the findings from this analysis will still provide supporting data to test the hypothesis.

The project that experienced conflict with the architect was associated with conflict with the client; it also took longer than the scheduled duration and resulted in a loss, this was the only project to be unsuccessful in all of the performance criteria. With regard to the relationship between positive and negative socio-emotional interaction and successful projects, the results shown in Table 8.3.3 add support to the previous discussion.

*Table 8.3.3* Results of positive and negative socio-emotional interaction, and giving and requesting task-based interaction: Projects that did and did not result in conflict with the architect

| Interaction categories | No conflict | | Conflict | | Total interaction observed | |
|---|---|---|---|---|---|---|
| | No. | % | No. | % | No. | % |
| Positive socio-emotional interaction | 1 051 | 8 | 138 | 7 | 1 189 | 8 |
| Giving task-based interaction | 9 502 | 73 | 1 637 | 79 | 11 139 | 74 |
| Requesting task-based interaction | 2 138 | 17 | 238 | 11 | 2 376 | 16 |
| Socio-emotional interaction | 300 | 2 | 73 | 4 | 373 | 3 |
| Total interaction observed | 12 991 | | 2 086 | | 15 077 | |

Notes
$\chi^2 = 51.232$, $df = 3$, $p = <0.001$.

Again, the amount of positive socio-emotional communication was significantly higher (8 per cent compared with 7 per cent) in projects that did not result in conflict. In this particular situation there was also a significantly higher level of negative socio-emotional communication used within the project that did experience conflict (4 per cent compared with 2 per cent). There was 6 per cent more positive than negative socio-emotional communication in the projects that did not result in conflict compared with a difference of 3 per cent in the project that did result in conflict.

Consistent with all previous analysis of successful project outcomes, the projects that did not experience conflict had a lower percentage of giving task-based communication acts than the projects that did result in conflict (73 per cent compared with 79 per cent). Also, consistent with the previous cost and duration analysis, projects that did not experience conflict used significantly higher levels of requesting task-based communication acts than projects that did (17 per cent compared with 11 per cent).

As in the duration analysis, comparisons between projects that did and did not experience conflict with the architect (Table 8.2.3) showed no significant difference in terms of their task and emotional communication acts. However, when the task-based categories were divided into giving and receiving task-based interaction and positive and negative socio-emotional acts, the difference was significant ($\chi^2 = 51.232$, $df = 3$, $\rho = <0.001$) (Table 8.3.3).

## Analysis of Bales' 12 categories

Although the line graphs in each of the comparisons of projects that were completed on time and within budget, or not (Figure 8.4.1), look similar, the differences between the profiles representing successful and unsuccessful outcomes are significant. At this level of analysis, all 12 communication acts are identified. When examining all 12 categories together it is difficult to identify trends and the interaction categories that are causing the differences. Following this analysis of all 12 acts, each of the IPA categories are analysed separately, helping to identify the individual detail of patterns associated with successful and unsuccessful project outcomes.

The difference between interaction of projects that were within and over budget produced a Pearson Chi-square value of ($\chi^2$) $= 122.534$, $df = 11$, $\rho = <0.001$. Projects that were completed within, and in excess of, the scheduled duration produced a value of $\chi^2 = 94.598$, $df = 11$, $\rho = <0.001$. The project associated with conflict with the architect was also found to be significantly different from the other projects that did not experience conflict ($\chi^2 = 161.010$, $df = 11$, $\rho = <0.001$). The results were consistent in showing that the overall differences found in the interaction associated with successful and unsuccessful projects were greater than would be attributable to chance (Table 8.4.1).

The results of the duration analysis are particularly interesting because the number of projects that were completed within, and in excess of, the scheduled duration were the same – five projects being successful and five unsuccessful. Due

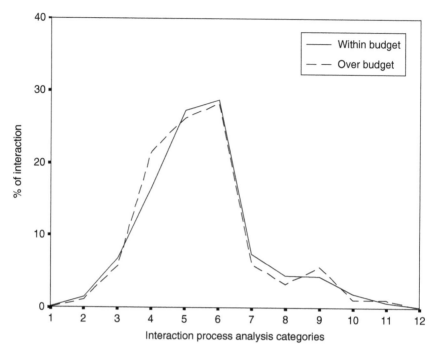

*Figure 8.4.1* IPA for group meetings of projects that were within budget and those that were over budget (projects over budget also experienced conflict between the client and the contractor)

to this occurrence it was possible to compare the total number of communication acts in each case. It is notable that the count for the number of communication acts for unsuccessful projects exceeded the communication acts of successful projects by 22 per cent (Table 8.4.2).

|  | *Total communication acts (15077)* |  | *% of total interaction* |
|---|---|---|---|
| Over schedule | 9 175 |  | 61 |
| Within schedule | 5 902 | less | 39 |
|  |  |  | 22 |

Participants working on the successful projects tended to interact less during meetings. Their communication was distributed across a wider range of communication acts compared to those working on the less successful projects (Figures 8.4.1, 8.4.2 and 8.4.3). Contributors to successful projects made greater use of more extreme emotional acts, as well as using higher levels of requesting task-based interaction. Those operating on less successful projects tended to use interaction that is concentrated around the giving task-based interaction categories. It is important to note that while less interaction took place during

*Table 8.4.1* Results for full IPA set: Projects completed within and over budget

| Interaction categories | Within budget | | Over budget | | Total interaction observed | |
|---|---|---|---|---|---|---|
| | *No.* | *%* | *No.* | *%* | *No.* | *%* |
| 1: Shows solidarity | 17 | 0 | 1 | 0 | 18 | 0 |
| 2: Shows tension release | 134 | 1 | 59 | 1 | 193 | 1 |
| 3: Agrees | 665 | 7 | 313 | 6 | 978 | 7 |
| 4: Gives suggestion | 1 612 | 17 | 1 164 | 22 | 2 776 | 18 |
| 5: Gives opinion | 2 645 | 27 | 1 415 | 26 | 4 060 | 27 |
| 6: Gives information | 2 780 | 29 | 1 523 | 28 | 4 303 | 29 |
| 7: Asks for orientation | 721 | 7 | 321 | 6 | 1 042 | 7 |
| 8: Asks for opinion | 432 | 5 | 177 | 3 | 609 | 4 |
| 9: Asks for suggestion | 422 | 4 | 303 | 6 | 725 | 5 |
| 10: Disagrees | 185 | 2 | 58 | 1 | 243 | 2 |
| 11: Shows tension | 66 | 1 | 56 | 1 | 122 | 1 |
| 12: Shows antagonism | 8 | 0 | 0 | 0 | 8 | 0 |
| Total interaction observed | 9 687 | | 5 390 | | 15 077 | |

Notes
$\chi^2 = 122.534$, $df = 11$, $p = <0.001$.

*Table 8.4.2* Results for full IPA set: Projects that were completed within and exceeded the scheduled duration

| Interaction categories | Within schedule | | Over schedule | | Total interaction observed | |
|---|---|---|---|---|---|---|
| | *No.* | *%* | *No.* | *%* | *No.* | *%* |
| 1: Shows solidarity | 15 | 0 | 3 | 0 | 18 | 0 |
| 2: Shows tension release | 94 | 2 | 99 | 1 | 193 | 1 |
| 3: Agrees | 396 | 7 | 582 | 6 | 978 | 7 |
| 4: Gives suggestion | 1 006 | 17 | 1 770 | 19 | 2 776 | 18 |
| 5: Gives opinion | 1 575 | 27 | 2 485 | 27 | 4 060 | 27 |
| 6: Gives information | 1 750 | 30 | 2 553 | 28 | 4 303 | 29 |
| 7: Asks for orientation | 474 | 8 | 568 | 6 | 1 042 | 7 |
| 8: Asks for opinion | 234 | 4 | 375 | 4 | 609 | 4 |
| 9: Asks for suggestion | 266 | 5 | 459 | 5 | 725 | 5 |
| 10: Disagrees | 66 | 1 | 177 | 2 | 243 | 2 |
| 11: Shows tension | 22 | 0 | 100 | 1 | 122 | 1 |
| 12: Shows antagonism | 4 | 0 | 4 | 0 | 8 | 0 |
| Total interaction observed | 5 902 | | 9 175 | | 15 077 | |

Notes
$\chi^2 = 94.598$, $df = 11$, $p = <0.001$.

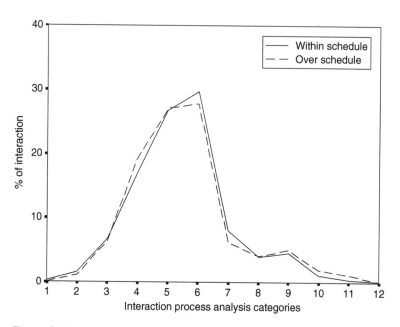

*Figure 8.4.2* IPA for group meetings of projects that were completed within the scheduled time and those that went over the scheduled duration

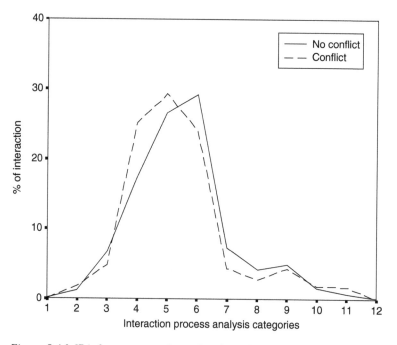

*Figure 8.4.3* IPA for group meetings of projects that were completed with and without conflict emerging between the contractor and the architect

*Table 8.4.3* Results for full IPA set: Projects that did and did not result in conflict with the architect

| Interaction categories | No conflict | | Conflict | | Total interaction observed | |
|---|---|---|---|---|---|---|
| | No. | % | No. | % | No. | % |
| 1: Shows solidarity | 18 | 0 | 0 | 0 | 18 | 0 |
| 2: Shows tension release | 156 | 1 | 37 | 2 | 193 | 1 |
| 3: Agrees | 877 | 7 | 101 | 5 | 978 | 7 |
| 4: Gives suggestion | 2 251 | 17 | 525 | 25 | 2 776 | 18 |
| 5: Gives opinion | 3 449 | 27 | 611 | 29 | 4 060 | 27 |
| 6: Gives information | 3 802 | 29 | 501 | 24 | 4 303 | 29 |
| 7: Asks for orientation | 951 | 7 | 91 | 4 | 1 042 | 7 |
| 8: Asks for opinion | 552 | 4 | 57 | 3 | 609 | 4 |
| 9: Asks for suggestion | 635 | 5 | 90 | 4 | 725 | 5 |
| 10: Disagrees | 206 | 2 | 37 | 2 | 243 | 2 |
| 11: Shows tension | 86 | 1 | 36 | 2 | 122 | 1 |
| 12: Shows antagonism | 8 | 0 | 0 | 0 | 8 | 0 |
| Total interaction observed | 12 991 | | 2 086 | | 15 077 | |

Notes
$\chi^2 = 161.010$, $df = 11$, $p = <0.001$.

the successful projects a much broader range of communication acts were used. The successful groups show a more developed pattern of interaction. The projects that did and did not result in conflict with the architect are tabulated in Table 8.4.3.

Considering the number of communication acts observed and the number of variables (12 communication acts in 2 groups), it is likely that a difference will be found when using the Chi-square test. It is possible that only some of the communication acts account for the significant difference shown. Although the adjusted residuals can be used to identify individual variables that are different from the expected results, they do not show the extent that the individual communication act accounts for the significant difference. The examination of categories in isolation will add support to the analysis of the full data set. The individual analysis reduces the number of variables examined. However, with multiple analysis the possibility of type-one errors is increased, thus a significance value of $p = 0.01$ is considered to be more reliable.

### Results: IPA category 1 – shows solidarity

Category 1 represents the following types of interaction: shows solidarity, raises others status, gives help or reward. A more detailed explanation of each IPA category, and other events and behaviours that are coded for each of the IPA categories, can be found in Appendix 1.

*Table 8.4.4* Results for IPA category 1: Projects that were within budget and projects that were over budget

| Interaction categories | Within budget | | Over budget | | Total interaction observed | |
|---|---|---|---|---|---|---|
| | No. | % | No. | % | No. | % |
| All other categories | 9670 | 100 | 5389 | 100 | 15059 | 100 |
| Shows solidarity | 17 | 0 | 1 | 0 | 18 | 0 |
| Total interaction observed | 9687 | | 5390 | | 15077 | |

Notes
Pearson Chi-square results $= \chi^2 = 7.153$, $df = 1$, $p = 0.007$.

Using the Chi-square statistic, a significant difference was found between projects that completed within and over budget. While representing less than 1 per cent of the total interaction, the higher levels of IPA category 1 were associated with projects that completed within budget.

*Table 8.4.5* Results for IPA category 1: Projects that were completed within the scheduled duration and projects that exceeded the scheduled duration

| Interaction categories | Within schedule | | Over schedule | | Total interaction observed | |
|---|---|---|---|---|---|---|
| | No. | % | No. | % | No. | % |
| All other categories | 5887 | 100 | 9172 | 100 | 15059 | 100 |
| Shows solidarity | 15 | 0 | 3 | 0 | 18 | 0 |
| Total interaction observed | 5902 | | 9175 | | 15077 | |

Notes
Pearson Chi-square results $= \chi^2 = 14.771$, $df = 1$, $p = <0.001$.

A significant difference was found in the analysis of scheduled duration. Higher levels of category 1 were associated with projects that completed on time.

*Table 8.4.6* Results for IPA category 1: Projects that did not experience conflict with the architect and that did experience conflict with the architect

| Interaction categories | No conflict | | Conflict | | Total interaction observed | |
|---|---|---|---|---|---|---|
| | No. | % | No. | % | No. | % |
| All other categories | 12973 | 100 | 2086 | 100 | 15059 | 100 |
| Shows solidarity | 18 | 0 | 0 | 0 | 18 | 0 |
| Total interaction observed | 12991 | | 2086 | | 15077 | |

Notes
Pearson Chi-square results $= \chi^2 = 2.894$, $df = 1$, $p = 0.089$.

No significant difference was found between the levels of category 1 that occurred in projects, which did or did not experience conflict with the architect. Although no significant difference was found, all of the communication acts that showed attributes of support and solidarity were found in the projects that did not experience conflict.

## Detailed analysis of category 1 – shows solidarity

The first observation of category 1 was that the levels of occurrence are very low in both positive and negative project outcomes. Although it was low in the successful project outcomes, it was even lower in the unsuccessful project outcomes and rarely occurred in unsuccessful cases.

The $\rho$ values of the Pearson Chi-square statistic shows significant differences in the levels of category 1 (agreeing) in the projects that were examined against the budget ($\chi^2 = 7.153$, $df = 1$, $\rho = 0.007$) and scheduled duration ($\chi^2 = 14.771$, $df = 1$, $\rho = <0.001$) criteria. Although the statistical analyses were undertaken, the Chi-square test is not accurate when cell counts for more than 20 per cent of the cells are less than five. In all cross-tabulations that examined showing solidarity (category 1), one of the cell counts associated with unsuccessful projects was less than five, representing 25 per cent of the cells.

The observations of category 1 were very small, the highest count being 15. In all of the studies, this category occurs less than 1 per cent of the time. Where it does occur, it was higher in situations that were associated with positive outcomes. The occurrence of this more extreme socio-emotional act, in cases with positive outcomes, may be indicating that these groups have established relationships that allow the M&D group to progress slightly further in group development. Alternatively, this could suggest that some participants in the successful groups use an extended range of communication behaviours regardless of the group development and this could be an attribute of the more successful communicator. Even though the successful projects used less communication acts, the communication of the successful group was distributed across more categories, with a few occurrences in the extreme emotional categories (Table 8.4.2). Due to the category's lack of use, when it did emerge in meetings it was very notable. The most extreme positive socio-emotional act being used to reinforce relationships and give praise to another person. Offering praise and support often occurred following episodes of conflict and was clearly used to maintain, strengthen or re-establish a firm relationship.

Showing solidarity is an extreme positive emotional act. When the act occurred it was used to show support, for example openly suggesting that another member had made a good decision. It was also used following tense negotiations, when a course of action had been established and members of the group showed strong support for the action. The expression shows faith in the decision or actions to be taken. It is interesting that such strong support had a tendency to occur in groups that ultimately resulted in greater levels of success. Those operating in successful projects are confident enough to show their support for decisions. It could be suggested that the successful groups had more reason to show such positive emotion, as the participant probably knew the project was performing well, although it is also worth noting that the successful and unsuccessful projects experienced similar amounts of negative emotion.

### Results: IPA category 2

IPA category 2 represents the following types of interaction: shows tension release, jokes, laughs, shows satisfaction.

*Table 8.4.7* Results for IPA category 2: Projects that were within budget and over budget

| Interaction categories | Within budget | | Over budget | | Total interaction observed | |
|---|---|---|---|---|---|---|
| | No. | % | No. | % | No. | % |
| All other categories | 9 553 | 99 | 5 331 | 99 | 14 884 | 99 |
| Shows tension release | 134 | 1 | 59 | 1 | 193 | 1 |
| Total interaction observed | 9 687 | | 5 390 | | 15 077 | |

Notes
Pearson Chi-square results = $\chi^2 = 2.284$, $df = 1$, $p = 0.131$.

No significant difference was found.

*Table 8.4.8* Results for IPA category 2: Projects that were completed within and exceeded the scheduled duration

| Interaction categories | Within schedule | | Over schedule | | Total interaction observed | |
|---|---|---|---|---|---|---|
| | No. | % | No. | % | No. | % |
| All other categories | 5 808 | 98 | 9 076 | 99 | 14 884 | 99 |
| Shows tension release | 94 | 2 | 99 | 1 | 193 | 1 |
| Total interaction observed | 5 902 | | 9 175 | | 15 077 | |

Notes
Pearson Chi-square results = $\chi^2 = 7.499$, $df = 1$, $p = 0.006$.

A significant difference was found in the analysis of scheduled duration. Higher levels of category 2 were associated with projects that completed on time.

*Table 8.4.9* Results for IPA category 2: Projects that did not and did experience conflict with the architect

| Interaction categories | No conflict | | Conflict | | Total interaction observed | |
|---|---|---|---|---|---|---|
| | No. | % | No. | % | No. | % |
| All other categories | 12 835 | 99 | 2 049 | 98 | 14 884 | 99 |
| Shows tension release | 156 | 1 | 37 | 2 | 193 | 1 |
| Total interaction observed | 12 991 | | 2 086 | | 15 077 | |

Notes
Pearson Chi-square results = $\chi^2 = 4.668$, $df = 1$, $p = 0.031$.

A significant difference was found (but only at the <0.05 level) in the analysis of conflict between the contractor and the client. Higher levels of category 2 were associated with projects that did experience conflict with the architect. However, due to the multiple analysis, such results are not a strong indicator of significance.

### Detailed analysis of category 2 – showing tension release

Although this category has a low level of occurrence, it was higher than category 1. The levels of interaction in this category were inconsistent. Using the Chi-square statistic, significant differences were found in the levels of IPA category 2 (showing tension release). The projects that were examined against the scheduled duration criteria ($\chi^2 = 7.499$, $df = 1$, $\rho = 0.006$) and those that assessed the occurrence of conflict between the contractor and the architect ($\chi^2 = 4.668$, $df = 1$, $\rho = 0.031$) produced significant results. However, the results were inconsistent, with higher levels of category 2 being associated with the successful outcome in the duration analysis and higher levels in unsuccessful outcomes in the conflict analysis. No significant difference was found in the projects that did and did not complete within budget ($\chi^2 = 2.284$, $df = 1$, $\rho = 0.131$).

Showing tension release often occurred after episodes of tension or while tension seemed to be developing. In some groups, tension was expressed and directed at an individual member or professional group (e.g. architects, mechanical and electrical consultants, contractors etc.) and sometimes it was expressed as bouts of frustration that had little or no real direction. When tension was given direction it was more likely to receive a response that resulted in action than when it was a mere expression of frustration; however, with both episodes of frustration or directed negative emotion it was common to see tension release following the expression of tension. The observations did not reveal any differences in the ways the groups expressed their tension release.

### Results: IPA category 3

Category 3 represents the following types of interaction: agrees, shows passive acceptance, understands, concurs, complies.

*Table 8.4.10* Results for IPA category 3: Projects that were within budget and projects that were over budget

| Interaction categories | Within budget | | Over budget | | Total interaction observed | |
|---|---|---|---|---|---|---|
| | No. | % | No. | % | No. | % |
| All other categories | 9022 | 93 | 5077 | 94 | 14099 | 94 |
| Agrees | 665 | 7 | 313 | 6 | 978 | 7 |
| Total interaction observed | 9687 | | 5390 | | 15077 | |

Notes
Pearson Chi-square results $= \chi^2 = 6.388$, $df = 1$, $\rho = 0.011$.

A significant difference was found here. Category 3 has a significantly higher level of occurrence in projects that completed within budget, although due to the level of multiple analysis this was not a strong indicator of significance.

*Table 8.4.11* Results for IPA category 3: Projects that were completed within and exceeded the scheduled duration

| Interaction categories | Within schedule | | Over schedule | | Total interaction observed | |
|---|---|---|---|---|---|---|
| | No. | % | No. | % | No. | % |
| All other categories | 5 506 | 93 | 8 593 | 94 | 14 099 | 94 |
| Agrees | 396 | 7 | 582 | 6 | 978 | 7 |
| Total interaction observed | 5 902 | | 9 175 | | 15 077 | |

Notes
Pearson Chi-square results = $\chi^2 = 0.794$, $df = 1$, $p = 0.373$.

No significant difference was found.

*Table 8.4.12* Results for IPA category 3: Projects that did not and did experience conflict with the architect

| Interaction categories | No conflict | | Conflict | | Total interaction observed | |
|---|---|---|---|---|---|---|
| | No. | % | No. | % | No. | % |
| All other categories | 12 114 | 93 | 1 985 | 95 | 14 099 | 94 |
| Agrees | 877 | 7 | 101 | 5 | 978 | 7 |
| Total interaction observed | 12 991 | | 2 086 | | 15 077 | |

Notes
Pearson Chi-square results = $\chi^2 = 10.799$, $df = 1$, $p = 0.001$.

A significant difference was found in the analysis of conflict between the contractor and the client. Higher levels of category 3 were associated with projects that did not experience conflict with the architect.

### Detailed analysis of category 3 – agrees

IPA category 3 occurred considerably more than did socio-emotional categories 1 and 2. All of the analyses indicated that category 3 occurred more frequently in the projects with successful outcomes. Using the Chi-square statistic, significant differences were found in the levels of IPA category 3 (agreeing) in the projects that were examined against the budget ($\chi^2 = 6.388$, $df = 1$, $\rho = 0.011$) and conflict between the contractor and the architect ($\chi^2 = 10.799$, $df = 1$, $\rho = 0.001$) criteria. No significant difference was found in the projects that did and did not complete within the scheduled duration ($\chi^2 = 0.794$, $df = 1$, $\rho = 0.373$).

Agreement was used in many ways, for example agreeing on an opinion, direction or suggestion. Unlike the more extreme positive acts, it did not show strong support for a proposal, but provided an indication that others were accepting the general idea. On a number of occasions the use of agreement would trigger a disagreement act. Often information and opinions could be presented without members agreeing or disagreeing, general ideas could be discussed without others indicating whether they believed they were right or wrong. However, if members were showing that they were agreeing with a suggestion, those members not agreeing had to quickly show their dissent or allow the proposal to be agreed. Occasionally members would refer back to a previous issue and reopen the discussion to ask more questions or show disagreement, but it was more common for disagreement to emerge when ideas were proposed and members showed support.

## Results: IPA category 4

Category four represents the following types of interaction: giving direction and implying autonomy for others.

*Table 8.4.13* Results for IPA category 4: Projects that were within and over budget

| Interaction categories | Within budget | | Over budget | | Total interaction observed | |
|---|---|---|---|---|---|---|
| | *No.* | *%* | *No.* | *%* | *No.* | *%* |
| All other categories | 8075 | 83 | 4226 | 78 | 12301 | 82 |
| Gives suggestion | 1612 | 17 | 1164 | 22 | 2776 | 18 |
| Total interaction observed | 9687 | | 5390 | | 15077 | |

Notes
Pearson Chi-square results $= \chi^2 = 56.593$, $df = 1$, $\rho = <0.001$.

A significant difference was found here. Category 4 has a higher level of occurrence in projects that were over budget.

*Table 8.4.14* Results for IPA category 4: Projects within over schedule

| Interaction categories | Within schedule | | Over schedule | | Total interaction observed | |
|---|---|---|---|---|---|---|
| | No. | % | No. | % | No. | % |
| All other categories | 4896 | 83 | 7405 | 81 | 12301 | 82 |
| Gives suggestion | 1006 | 17 | 1770 | 19 | 2776 | 18 |
| Total interaction observed | 5902 | | 9175 | | 15077 | |

Notes
Pearson Chi-square results $= \chi^2 = 12.066$, $df = 1$, $\rho = 0.001$.

A significant difference was found here. Category 4 occurs more in the projects that exceeded the scheduled duration.

*Table 8.4.15* Results for IPA category 4: Projects that did and did not experience conflict with the architect

| Interaction categories | No conflict | | Conflict | | Total interaction observed | |
|---|---|---|---|---|---|---|
| | No. | % | No. | % | No. | % |
| All other categories | 10740 | 83 | 1561 | 75 | 12301 | 82 |
| Gives suggestion | 2251 | 17 | 525 | 25 | 2776 | 18 |
| Total interaction observed | 12991 | | 2086 | | 15077 | |

Notes
Pearson Chi-square results $= \chi^2 = 73.551$, $df = 1$, $\rho = <0.001$.

A significant difference was found in the analysis of conflict between the contractor and the client. Higher levels of category 4 were associated with projects that did experience conflict with the architect.

### Detailed analysis of category 4 – gives suggestion

Significant differences were found in the levels of IPA category 4 (gives suggestions) in all of the analyses. Category four was consistently higher in projects with unsuccessful outcomes; budget ($\chi^2 = 56.593$, $df = 1$, $\rho = <0.001$), schedule duration ($\chi^2 = 12.066$, $df = 1$, $\rho = 0.001$) and conflict with the architect ($\chi^2 = 73.551$, $df = 1$, $\rho = <0.001$).

Category 4 was often used to offer direction to the group. Some of the members who adopted leading positions in the groups would be more prominent in offering suggested courses of action. Such statements would often be made after considerable information and opinions of problems had been presented. Once one person made a suggestion it was common to observe others offering slightly different suggestions. On many occasions when members offered direction to others there was an occurrence of agreement, disagreement and conflict. The use of suggestions was often used to bring discussions to a close. Attempting to close discussions in this way encouraged others to express their support, identify weaknesses with the proposal, or propose alternative solutions, rather than continuing to present ideas, information or opinions on the topic. In successful projects the use of suggestions were used to provide direction and guidance and resulted in action. In the less successful projects, suggestions were often vague and associated with spurious information. Even though many suggestions were made in the unsuccessful groups, they frequently lacked direction and referred to too many different aspects of the project, making discussion unnecessarily long. There is a difference in the way that suggestions and directions are given. Suggestions that are made with some authority, are precise, concise and clear, seem to have a greater impact on the group than suggestions which lack relevance to the action point and come from all members of the group regardless of their expertise. When too many members become engaged in discussion for long periods of time the meeting loses its direction.

### Results: IPA category 5

Category 5 represents gives opinion, evaluation, analysis, expresses a feeling.

*Table 8.4.16* Results for IPA category 5: Projects within budget and projects that were over budget

| Interaction categories | Within budget | | Over budget | | Total interaction observed | |
|---|---|---|---|---|---|---|
| | No. | % | No. | % | No. | % |
| All other categories | 7042 | 73 | 3975 | 74 | 11017 | 73 |
| Gives opinion | 2645 | 27 | 1415 | 26 | 4060 | 27 |
| Total interaction observed | 9687 | | 5390 | | 15077 | |

Notes
Pearson Chi-square results $= \chi^2 = 1.949$, $df = 1$, $p = 0.163$.

No significant difference was found.

*Table 8.4.17* Results for IPA category 5: Projects completed within the scheduled duration and projects that exceeded the scheduled duration

| Interaction categories | Within schedule | | Over schedule | | Total interaction observed | |
|---|---|---|---|---|---|---|
| | No. | % | No. | % | No. | % |
| All other categories | 4327 | 73 | 6690 | 73 | 11017 | 73 |
| Gives opinion | 1575 | 27 | 2485 | 27 | 4060 | 27 |
| Total interaction observed | 5902 | | 9175 | | 15077 | |

Notes
Pearson Chi-square results $= \chi^2 = 0.290$, $df = 1$, $p = 0.590$.

No significant difference was found.

*Table 8.4.18* Results for IPA category 5: Projects that did and did not experience conflict with the architect

| Interaction categories | No conflict | | Conflict | | Total interaction observed | |
|---|---|---|---|---|---|---|
| | No. | % | No. | % | No. | % |
| All other categories | 9 542 | 74 | 1 475 | 71 | 11 017 | 73 |
| Gives opinion | 3 449 | 27 | 611 | 29 | 4 060 | 27 |
| Total interaction observed | 12 991 | | 2 086 | | 15 077 | |

Notes
Pearson Chi-square results $= \chi^2 = 6.865$, $df = 1$, $\rho = 0.009$.

A significant difference was found in the analysis of conflict between the contractor and the client. The results for the project that did and did not experience conflict with the client show higher levels of category 5 being associated with projects that did experience conflict with the architect.

### Detailed analysis of category 5 – giving opinion

The results for category five are inconsistent. Significant differences were found in the levels of IPA category 5 (giving opinion) in only one of the statistical tests – conflict with architect ($\chi^2 = 6.865$, $df = 1$, $\rho = 0.009$). The results of the other statistical tests were not significant (budget $\chi^2 = 1.949$, $df = 1$, $\rho = 0.165$ and schedule duration $\chi^2 = 0.290$, $df = 1$, $\rho = 0.59$).

Opinions and beliefs were often used to analyse problems and explore issues; however, the decisive group action would only develop once levels of agreement and disagreement emerged. Thus, information, opinions and analysis could be presented in discussions without the group deciding on a course of action.

### Results IPA category 6

Category 6 represents interaction associated with: gives orientation, information, repeats, clarifies, confirms.

*Table 8.4.19* Results for IPA category 6: Projects within budget and projects that were over budget

| Interaction categories | Within budget | | Over budget | | Total interaction observed | |
|---|---|---|---|---|---|---|
| | No. | % | No. | % | No. | % |
| All other categories | 6 907 | 71 | 3 867 | 72 | 10 774 | 71 |
| Gives information | 2 780 | 29 | 1 523 | 28 | 4 303 | 29 |
| Total interaction observed | 9 687 | | 5 390 | | 15 077 | |

Notes
Pearson Chi-square results $= \chi^2 = 0.332$, $df = 1$, $\rho = 0.564$.

No significant difference was found.

*Table 8.4.20* Results for IPA category 6: Projects completed within the scheduled duration and projects that exceeded the scheduled duration

| Interaction categories | Within schedule | | Over schedule | | Total interaction observed | |
|---|---|---|---|---|---|---|
| | No. | % | No. | % | No. | % |
| All other categories | 4 152 | 70 | 6 622 | 72 | 10 774 | 72 |
| Gives information | 1 750 | 30 | 2 553 | 28 | 4 303 | 29 |
| Total interaction observed | 5 902 | | 9 175 | | 15 077 | |

Notes
Pearson Chi-square results $= \chi^2 = 5.868$, $df = 1$, $p = 0.015$.

A significant difference was found, although due to the level of multiple analysis, this was not a strong indicator of significance. Giving information category 6 was higher in the projects that completed within the scheduled duration.

*Table 8.4.21* Results for IPA category 6: Projects that did and did not experience conflict with the architect

| Interaction categories | No conflict | | Conflict | | Total interaction observed | |
|---|---|---|---|---|---|---|
| | No. | % | No. | % | No. | % |
| All other categories | 9 189 | 71 | 1 585 | 76 | 10 774 | 72 |
| Gives information | 3 802 | 29 | 501 | 24 | 4 303 | 29 |
| Total interaction observed | 12 991 | | 2 086 | | 15 077 | |

Notes
Pearson Chi-square results $= \chi^2 = 24.283$, $df = 1$, $p = <0.001$.

A significant difference was found in the analysis of conflict between the contractor and the client. Higher levels of category 6 are associated with projects that did not experience conflict with the architect.

### Detailed analysis of category 6 – giving orientation

The results of this category are consistently higher in cases with positive outcomes. Using the Chi-square statistic, significant differences were found in the levels of IPA category 6 (giving information) in two of the analyses: scheduled duration ($\chi^2 = 5.868$, $df = 1$, $p = 0.015$) and conflict with architect ($\chi^2 = 24.283$, $df = 1$, $p = <0.001$). All of the results in this set of analyses showed higher frequencies of category 6 in the positive outcomes. Although the budget analysis also showed higher levels of category 6 in the positive outcome, it was not significant ($\chi^2 = 0.332$, $df = 1$, $p = 0.564$). Giving orientation was used to present the information and facts relating to project issues. High levels of information were presented to aid understanding and help others follow the issues associated with the problems being discussed.

### Results: IPA category 7

Category 7 represents interaction such as: asks for orientation, information, repetition, confirmation.

*Table 8.4.22* Results for IPA category 7: Projects within budget and projects that were over budget

| Interaction categories | Within budget | | Over budget | | Total interaction observed | |
|---|---|---|---|---|---|---|
| | No. | % | No. | % | No. | % |
| All other categories | 8966 | 93 | 5069 | 94 | 14035 | 93 |
| Asks for orientation | 721 | 7 | 321 | 6 | 1042 | 7 |
| Total interaction observed | 9687 | | 5390 | | 15077 | |

Notes
Pearson Chi-square results $= \chi^2 = 11.910$, $df = 1$, $p = 0.001$.

A significant difference was found. Higher frequencies of the communication act asking for orientation were found in the projects with a successful budget outcome.

*Table 8.4.23* Results for IPA category 7: Projects completed within the scheduled duration, and projects that exceeded the scheduled duration

| Interaction categories | Within schedule | | Over schedule | | Total interaction observed | |
|---|---|---|---|---|---|---|
| | No. | % | No. | % | No. | % |
| All other categories | 5428 | 92 | 8607 | 94 | 14035 | 93 |
| Asks for orientation | 474 | 8 | 568 | 6 | 1042 | 7 |
| Total interaction observed | 5902 | | 9175 | | 15077 | |

Notes
Pearson Chi-square results $= \chi^2 = 18.910$, $df = 1$, $p = {<}0.001$.

A significant difference was found. Asking for orientation category 7 was higher in the projects that completed within the scheduled duration.

*Table 8.4.24* Results for IPA category 7: Projects that did and did not experience conflict with the architect

| Interaction categories | No conflict | | Conflict | | Total interaction observed | |
|---|---|---|---|---|---|---|
| | No. | % | No. | % | No. | % |
| All other categories | 12 040 | 93 | 1 995 | 96 | 14 035 | 93 |
| Asks for orientation | 951 | 7 | 91 | 4 | 1 042 | 7 |
| Total interaction observed | 12 991 | | 2 086 | | 15 077 | |

Notes
Pearson Chi-square results $= \chi^2 = 24.445$, $df = 1$, $p = <0.001$.

A significant difference was found in the use of communication act 7 when examined against projects that did and did not experience conflict between the contractor and the architect. The results show higher levels of category 7 being associated with projects that did not experience conflict.

### Detailed analysis of category 7 – asking for orientation

This category was consistent across all examinations, occurring more in the positive outcomes. Using the Chi-square statistic, significant differences were found in the levels of IPA category 7 (asking for orientation) in all three statistical tests: budget ($\chi^2 = 11.910$, $df = 1$, $p = 0.001$), schedule duration ($\chi^2 = 18.910$, $df = 1$, $p = <0.001$) and conflict ($\chi^2 = 24.445$, $df = 1$, $p = <0.001$). All of the results in this set of analyses have higher frequencies of category 7 in the positive outcomes and in all cases the results are significant. Asking for information was often used to delve further into issues. Where participants presented scant contextual or background information on a problem, other group members would occasionally ask for further information. Surprisingly, even in the multi-disciplinary group, different specialists would not openly admit to not understanding specialised information. To overcome difficulties presented by specialised issues, some group participants would request more detailed and fundamental background information.

### Results: IPA category 8

Category 8 represents communication acts such as: asks for opinion, evaluation, analysis and expression of feeling.

*Table 8.4.25* Results for IPA category 8: Projects within budget and projects that were over budget

| Interaction categories | Within budget | | Over budget | | Total interaction observed | |
|---|---|---|---|---|---|---|
| | No. | % | No. | % | No. | % |
| All other categories | 9 255 | 96 | 5 213 | 97 | 14 468 | 96 |
| Asks for opinion | 432 | 5 | 177 | 3 | 609 | 4 |
| Total interaction observed | 9 687 | | 5 390 | | 15 077 | |

Notes
Pearson Chi-square results $= \chi^2 = 12.350$, $df = 1$, $p = <0.001$.

A significant difference was found. Higher frequencies of the communication act asking for opinion and evaluation were found in the projects with a successful budget outcome.

*Table 8.4.26* Results for IPA category 8: Projects completed within and that exceeded the scheduled duration

| Interaction categories | Within schedule | | Over schedule | | Total interaction observed | |
|---|---|---|---|---|---|---|
| | No. | % | No. | % | No. | % |
| All other categories | 5 668 | 96 | 8 800 | 96 | 14 468 | 96 |
| Asks for opinion | 234 | 4 | 375 | 4 | 609 | 4 |
| Total interaction observed | 5 902 | | 9 175 | | 15 077 | |

Notes
Pearson Chi-square results $= \chi^2 = 139$, $df = 1$, $p = 0.709$.

No significant difference was found.

*Table 8.4.27* Results for IPA category 8: Projects that did and did not experience conflict with the architect

| Interaction categories | No conflict | | Conflict | | Total interaction observed | |
|---|---|---|---|---|---|---|
| | No. | % | No. | % | No. | % |
| All other categories | 12 439 | 96 | 2 029 | 97 | 14 468 | 96 |
| Asks for opinion | 552 | 4 | 57 | 3 | 609 | 4 |
| Total interaction observed | 12 991 | | 2 086 | | 15 077 | |

Notes
Pearson Chi-square results $= \chi^2 = 10.666$, $df = 1$, $p = 0.001$.

A significant difference was found in the use of category 8 for projects that did and did not experience conflict between the contractor and the client. The cross-tabulation shows higher levels of category 8 being associated with projects that did not experience conflict with the architect.

## Detailed analysis of category 8 – asking for opinion

The results in this category have a degree of inconsistency. Using the Chi-square statistic, significant differences were found in two of the results: budget ($\chi^2 = 12.350$, $df = 1$, $\rho = <0.001$) and conflict with the architect ($\chi^2 = 10.666$, $df = 1$, $\rho = 0.001$). The two sets of results have higher frequencies of category 8 in the positive outcomes. The analysis of projects that did and did not complete within the scheduled duration were not significant ($\chi^2 = 139$, $df = 1$, $\rho = 0.709$), although the unsuccessful projects had a marginally higher level of category 8.

Asking for opinion, explanation and analysis was used to ensure that information was not presented without a thorough explanation. This encouraged members to defend their suggestions, explaining how they would be implemented, how obstacles would be overcome and how other related issues would be dealt with. Such questioning often exposed the weakness of a suggestion preventing under-developed proposals being discussed further.

## Results: IPA category 9

Category 9 represents communication acts such as: ask for suggestion, direction, possible ways of action.

*Table 8.4.28* Results for IPA category 9: Projects within budget and projects that were over budget

| Interaction categories | Within budget | | Over budget | | Total interaction observed | |
|---|---|---|---|---|---|---|
| | No. | % | No. | % | No. | % |
| All other categories | 9265 | 96 | 5087 | 94 | 14352 | 95 |
| Asks for suggestion | 422 | 4 | 303 | 6 | 725 | 5 |
| Total interaction observed | 9687 | | 5390 | | 15077 | |

Notes
Pearson Chi-square results $= \chi^2 = 12.110$, $df = 1$, $\rho = 0.001$.

A significant difference was found. Higher frequencies of the communication act asking for suggestions were found in the projects that were over budget.

*Table 8.4.29* Results for IPA category 9: Projects completed within and over scheduled duration

| Interaction categories | Within schedule | | Over schedule | | Total interaction observed | |
|---|---|---|---|---|---|---|
| | *No.* | *%* | *No.* | *%* | *No.* | *%* |
| All other categories | 5 636 | 96 | 8 716 | 95 | 14 352 | 95 |
| Asks for suggestion | 266 | 5 | 459 | 5 | 725 | 5 |
| Total interaction observed | 5 902 | | 9 175 | | 15 077 | |

Notes
Pearson Chi-square results $= \chi^2 = 1.929$, $df = 1$, $\rho = 0.165$.

No significant difference was found.

*Table 8.4.30* Results for IPA category 9: Projects that did and did not experience conflict with the architect

| Interaction categories | No conflict | | Conflict | | Total interaction observed | |
|---|---|---|---|---|---|---|
| | *No.* | *%* | *No.* | *%* | *No.* | *%* |
| All other categories | 12 356 | 95 | 1 996 | 96 | 14 352 | 95 |
| Asks for suggestion | 635 | 5 | 90 | 4 | 725 | 5 |
| Total interaction observed | 12 991 | | 2 086 | | 15 077 | |

Notes
Pearson Chi-square results $= \chi^2 = 1.292$, $df = 1$, $\rho = 0.255$.

No significant difference was found.

### Detailed analysis of category 9 – asks for suggestion

The results for this category are inconsistent. Significant differences were found in the budget analysis only: budget ($\chi^2 = 12.110$, $df = 1$, $\rho = 0.001$), scheduled duration ($\chi^2 = 1.929$, $df = 1$, $\rho = 0.165$) and conflict with the architect ($\chi^2 = 1.292$, $df = 1$, $\rho = 0.255$). Where a significant difference was found, the level of communication act nine was higher in the unsuccessful outcome. In the other two sets of results, although not significant, the communication act nine was higher in the projects that were over schedule and higher in projects that did not experience conflict.

Asking for suggestions was used mostly to encourage others who had initially given information, ideas and opinions on an issue to turn them into a firm proposal. Occasionally they were rather hard statements insisting that others commit their resources and offer a solution.

### Results: IPA category 10

Category 10 represents the following communication acts: disagrees, shows passive rejection, formality, withholds help.

*Table 8.4.31* Results for IPA category 10: Projects within budget and projects that were over budget

| Interaction categories | Within budget | | Over budget | | Total interaction observed | |
|---|---|---|---|---|---|---|
| | *No.* | *%* | *No.* | *%* | *No.* | *%* |
| All other categories | 9 502 | 98 | 5 332 | 99 | 14 834 | 98 |
| Disagrees | 185 | 2 | 58 | 1 | 243 | 2 |
| Total interaction observed | 9 687 | | 5 390 | | 15 077 | |

Notes
Pearson Chi-square results $= \chi^2 = 15.180$, $df = 1$, $p = <0.001$.

A significant difference was found. Higher frequencies of the communication act disagreeing were found in the projects that were within budget.

*Table 8.4.32* Results for IPA category 10: Projects that completed within and that exceeded the scheduled duration

| Interaction categories | Within schedule | | Over schedule | | Total interaction observed | |
|---|---|---|---|---|---|---|
| | *No.* | *%* | *No.* | *%* | *No.* | *%* |
| All other categories | 5 836 | 99 | 8 998 | 98 | 14 834 | 98 |
| Disagrees | 66 | 1 | 177 | 2 | 243 | 2 |
| Total interaction observed | 5 902 | | 9 175 | | 15 077 | |

Notes
Pearson Chi-square results $= \chi^2 = 14.893$, $df = 1$, $p = <0.001$.

A significant difference was found. Higher levels of disagreement were found in projects that did not complete within schedule.

*Table 8.4.33* Results for IPA category 10: Projects that did and did not experience conflict with the architect

| Interaction categories | No conflict | | Conflict | | Total interaction observed | |
|---|---|---|---|---|---|---|
| | *No.* | *%* | *No.* | *%* | *No.* | *%* |
| All other categories | 12 785 | 98 | 2 049 | 98 | 14 834 | 98 |
| Disagrees | 206 | 2 | 37 | 2 | 243 | 2 |
| Total interaction observed | 12 991 | | 2 086 | | 15 077 | |

Notes
Pearson Chi-square results $= \chi^2 = 0.401$, $df = 1$, $\rho = 0.527$.

No significant difference was found.

### Detailed analysis of category 10 – disagrees

The results for this category are inconsistent. Using the Chi-square statistic, significant differences were found in only two of the results: budget ($\chi^2 = 15.180$, $df = 1$, $\rho = < 0.001$) and time ($\chi^2 = 14.893$, $df = 1$, $\rho = < 0.001$); however, the two results conflicted. Projects that were completed within budget exhibited higher levels of category 10, and projects that exceeded the scheduled duration were also associated with higher levels of category 10. Interaction that did or did not experience conflict with the architect was not significantly different ($\chi^2 = 0.401$, $df = 1$, $\rho = 0.527$).

Conflict emerged as a result of simple disagreements over opinions, statements and proposals, or as a result of different personal or organisational interests. Very few members of the M&D team would engage in conflict. Even after a short period of observation, it was notable who was most likely to disagree with others. Thus, it did not take long for the observer to become aware of the group's socio-emotional structure. Following disagreements, and thorough evaluation of ideas, courses of action were often proposed and agreed. Disagreeing with others helped expose and resolve technical and organisational problems.

### Results: IPA category 11

Category 11 represents the following communication acts: shows tension, asks for help and withdraws out of field.

*Table 8.4.34* Results for IPA category eleven: Projects that completed within and over budget

| Interaction categories | Within budget | | Over budget | | Total interaction observed | |
|---|---|---|---|---|---|---|
| | No. | % | No. | % | No. | % |
| All other categories | 9 621 | 99 | 5 334 | 99 | 14 955 | 99 |
| Shows tension | 66 | 1 | 56 | 1 | 122 | 1 |
| Total interaction observed | 9 687 | | 5 390 | | 15 077 | |

Notes
Pearson Chi-square results $= \chi^2 = 5.519$, $df = 1$, $p = 0.019$.

A significant difference was found, although due to the level of multiple analysis this is not a strong indicator of significance. Higher frequencies of the communication act showing tension were found in the projects that were over budget.

*Table 8.4.35* Results for IPA category eleven: Projects that completed within and over schedule

| Interaction categories | Within schedule | | Over schedule | | Total interaction observed | |
|---|---|---|---|---|---|---|
| | No. | % | No. | % | No. | % |
| All other categories | 5 880 | 100 | 9 075 | 99 | 14 955 | 99 |
| Shows tension | 22 | 0 | 100 | 1 | 122 | 1 |
| Total interaction observed | 5 902 | | 9 175 | | 15 077 | |

Notes
Pearson Chi-square results $= \chi^2 = 23.015$, $df = 1$, $p = <0.001$.

A significant difference was found. Higher levels of tension were found in projects that did not complete within schedule.

*Table 8.4.36* Results for IPA category eleven: Projects that did and did not experience conflict with the architect

| Interaction categories | No conflict | | Conflict | | Total interaction observed | |
|---|---|---|---|---|---|---|
| | No. | % | No. | % | No. | % |
| All other categories | 12 905 | 99 | 2 050 | 98 | 14 955 | 99 |
| Shows tension | 86 | 1 | 36 | 2 | 122 | 1 |
| Total interaction observed | 12 991 | | 2 086 | | 15 077 | |

Notes
Pearson Chi-square results $= \chi^2 = 25.342$, $df = 1$, $p = <0.001$.

A significant difference was found. Higher levels of the showing tension communication act were associated with projects that resulted in conflict between the architect and the contractor.

### Detailed analysis of category 11 – showing tension

All of the results are consistent and indicate that the communication act 11 (shows tension) has a stronger association with negative project outcomes. Using the Chi-square statistic, significant differences were found in all of the results: budget ($\chi^2 = 5.519$, $df = 1$, $p = 0.019$) (although the budget result was not a strong indicator of significance), scheduled duration ($\chi^2 = 23.015$, $df = 1$, $p = < 0.001$) and conflict with the architect ($\chi^2 = 25.342$, $df = 1$, $p = <0.001$).

Unless directed specifically at another group member, the act of showing tension did not result in others offering to help or more thorough discussions of the problem. When members showed that they were frustrated with a problem or process, others would not offer to defend their behaviour or position, but would often remain silent until the party venting their frustration finished talking. However, when members were engaged in critical discussions with other members, and showed that they were becoming increasingly concerned about the other's view or proposal, the level of evaluation and critical interaction increased. As tension increased, the members' efforts towards resolving the problem increased.

### Results: IPA category 12

Category 12 represents the following communication acts: shows antagonism, deflates others status, defends or asserts self.

*Table 8.4.37* Results for IPA category 12: Projects within budget and projects that were over budget

| Interaction categories | Within budget | | Over budget | | Total interaction observed | |
|---|---|---|---|---|---|---|
| | *No.* | *%* | *No.* | *%* | *No.* | *%* |
| All other categories | 9 679 | 100 | 5 390 | 100 | 15 069 | 100 |
| Shows antagonism | 8 | 0 | 0 | 0 | 8 | 0 |
| Total interaction observed | 9 687 | | 5 390 | | 15 077 | |

Notes
Pearson Chi-square = $\chi^2 = 4.454$, $df = 1$, $p = 0.035$.

A significant difference was found (Table 10.4.37). Although the occurrence of this category was very low, higher frequencies were found in the projects that were within budget. Where counts in cells are less than five the reliability of the Chi-square statistic is reduced.

*Table 8.4.38* Results for IPA category 12: Projects completed within the scheduled duration and projects that exceeded the scheduled duration

| Interaction categories | Within schedule | | Over schedule | | Total interaction observed | |
|---|---|---|---|---|---|---|
| | *No.* | *%* | *No.* | *%* | *No.* | *%* |
| All other categories | 5898 | 100 | 9171 | 100 | 15069 | 100 |
| Shows antagonism | 4 | 0 | 4 | 0 | 8 | 0 |
| Total interaction observed | 5902 | | 9175 | | 15077 | |

Notes
Pearson Chi-square = $\chi^2 = 0.396$, $df = 1$, $p = 0.529$.

No significant difference was found.

*Table 8.4.39* Results for IPA category 12: Projects that did and did not experience conflict with the architect

| Interaction categories | No conflict | | Conflict | | Total interaction observed | |
|---|---|---|---|---|---|---|
| | *No.* | *%* | *No.* | *%* | *No.* | *%* |
| All other categories | 12983 | 100 | 2086 | 100 | 15069 | 100 |
| Shows antagonism | 8 | 0 | 0 | 0 | 8 | 0 |
| Total interaction observed | 12991 | | 2086 | | 15077 | |

Notes
Pearson Chi-square = $\chi^2 = 1.285$, $df = 1$, $p = 0.257$.

No significant difference was found.

### Detailed analysis of category 12 – shows antagonism

Significant differences were found in only one set of category 12 interaction results (budget). The occurrence of this category was very low (it is not possible to rely on the Chi-square statistic in this situation). While all of the results indicate that the communication act 12 (shows antagonism) occurs mainly in groups associated with positive project outcomes, the frequency observed was so low that it is difficult to draw meaningful conclusions.

# Successful and unsuccessful project outcomes: Relationships between categories

### Showing solidarity and showing antagonism

In all three analyses (Tables 8.4.4, 8.4.5, 8.4.6), budget, duration and conflict with the architect, category 1 (showing solidarity) and category 12 (showing antagonism) account for even less than 0.1 per cent of the communication in unsuccessful outcomes. The levels of category 1 are low in both the positive and the negative outcomes; however, the fact that they occur more in the positive outcomes may have some association with group development. The groups associated with successful outcomes may have developed relationships that allow them to use marginally higher amounts of very expressive socio-emotional acts. However, the occurrences of these categories are so low that reliable statistical inference cannot be drawn.

### Showing tension and showing tension release

The levels of interaction between positive and negative outcomes for category 2 (releasing tension) are not consistent across all analyses. However, category 11 (showing tension) is consistently and significantly higher in all of the categories associated with unsuccessful outcomes. Thus, higher levels of tension are not associated with successful outcomes. The level of frustration shown by some members did suggest that there was a loss of control. The occurrence of tension in unsuccessful projects could be an indication that group members are recognising problems that could potentially affect the project's performance.

The use of positive emotional acts (when grouped together) tended to have a stronger association with positive project outcomes. Positive emotional acts used to relieve tension occurred more following periods of critical debate, rather than when members showed they were frustrated with the project. Tension occurred in both groups, although those working in teams that were considered successful used greater amounts of positive emotional expression. Such acts help rebuild and maintain relationships.

### Agreeing and disagreeing

Category 3 (agreeing) was higher in successful cases; however, disagreeing (category 10) varies; it occurs less in successful schedule duration (significant difference) and more in the positive budget analysis (significant difference), but the analysis of conflict with the architect was not significant. In many of the projects the use of agreement would be particularly prominent after disagreement and critical discussions. The use of agreement and disagreement tended to intensify the level of discussion. Early emergence of agreement following a proposal would prompt others not in agreement to identify their concerns. Disagreeing encouraged others to defend their proposals.

### *Giving suggestions and requesting suggestions*

While giving suggestions (category 4) is used more in the projects associated with unsuccessful outcomes, requesting suggestions (category 9) is not consistent throughout the analyses. The budget analysis produced the only significant result, being higher in projects that were over budget. Both giving suggestions and requesting suggestions were used to direct the group. Even when asking for suggestions, the member requesting the action often adopted an autonomous position demanding that the other person suggest a course of action. The high use of these categories by different members of the same group suggests that group members are attempting to offer many different directions rather than follow one course of action. Too many suggestions show a lack of agreement.

### *Giving opinions and requesting opinions*

Neither giving opinions (category 5) nor requesting opinions (category 8) produced a significantly consistent set of results. However, two analyses in category 8 did show a significant difference, with higher frequencies being associated with the projects that resulted in positive outcomes. Both types of communication act were used to present and draw greater contextual information into the meeting forum. As well as providing opinions, the category was also used to evaluate and analyse statements presented. It was evident that there was often more detail to the problem than was initially put forward. Asking for greater explanation served to expose hidden issues and help understand technical requirements.

### *Giving orientation and requesting orientation*

Both giving orientation (category 6) and requesting orientation (category 7) occurred more in the cases with successful project outcomes, being significantly different from unsuccessful projects. This suggests that asking for orientation and information, and giving orientation and information are associated with the successful project teams. Categories 5, 6, 7 and 8 were often used together during sequences of orientation and evaluation. As members explored issues associated with a problem, information, opinion, explanation and evaluation were presented and requested, thus increasing the potential to develop greater understanding of facts, and other members' perspective of associated issues.

## Contractor's representatives and interpersonal interaction

The final hypothesis predicted that *the contractor's representatives considered to be the most effective exhibit significantly different interaction patterns from the contractor's representatives perceived to be less effective.*

Before progressing to discussion of the results it is important to reiterate how the contractor's representatives were rated, and on what information the results were based. Each of the contractor's representatives was awarded a performance value, based on their ability to repeatedly contribute to successful project outcomes. The managing directors provided the perceived value of effectiveness, using historic data from previous projects. A value of 1 indicates contractor's representatives who have a greater association with projects that resulted in a profitable outcome. A value of 5 shows a greater association with projects that were not as profitable. A rating of 1 identifies the professionals who are most effective and 5 indicates professionals who are satisfactory. As the construction industry has a reputation for being 'cut-throat', it was not anticipated that personnel would be retained if they were considered unsatisfactory, and this notion was supported as the managing directors failed to use the lowest rating (5). All of the contractor's personnel were rated above satisfactory level (4 to 1).

Although the graphs provide an overall trend, the tables are the most useful tool for investigating individual components, especially where all 12 interaction categories are analysed. The adjusted residuals are used to show which categories have the greatest difference. Those furthest away from their expected value (those with the highest positive or negative value) are those that are most significant and provide some indication of the categories that could be affecting the contractor's representatives' performance and behaviour.

The results presented in Tables 8.6.1 and 8.6.2 are interesting. Some of the contractor's representatives' individual IPA values have similarities to those found in the group results, being higher in categories that were associated with positive outcomes; however, some of the individual characteristics were in contrast to the group observations. The professionals rated as most effective have some different interaction tendencies, when compared with the groups' profiles that were associated with projects that completed within time, to budget and did not experience conflict.

In some of the IPA categories, there was a clear relationship with the perceived ability of the individual and the levels of interaction; the interaction of effective contractor's representatives being considerably different from those rated less effective. The following section identifies interaction characteristics that are consistent with contractor's representatives who were rated as most effective and least effective.

### *Task and socio-emotional analysis: Contractor's representatives*

Using the Pearson Chi-square test, the results of the four ratings of contractor's representatives were significantly different. The most effective contractor's representatives used greater amounts of positive and negative socio-emotional communication, higher levels of requesting task-based interaction and lower levels of giving task-based interaction, than the representatives perceived to be less effective.

Table 8.6.1 Results of task-based and socio-emotional communication acts against perceived effectiveness of contractor's representatives' performance

| Contractor rating | Most effective | | | | | | 3 | | | Least effective | | | Total | |
| | 1 | | | 2 | | | | | | 4 | | | | |
| Interaction category | No. | % | Adj. | No. | % | Adj. | No. | % | Adj. | No. | % | Adj. | No. | % |
|---|---|---|---|---|---|---|---|---|---|---|---|---|---|---|
| Positive socio-emotional | 130 | 11 | 6 | 200 | 6 | -3.5 | 13 | 6 | -0.5 | 27 | 5 | -2.1 | 370 | 7 |
| Giving task-based | 730 | 62 | -10.9 | 2447 | 76 | 4.2 | 180 | 88 | 4.7 | 456 | 84 | 5.3 | 3813 | 74 |
| Requesting task-based | 277 | 24 | 8.5 | 455 | 14 | -3.7 | 11 | 5 | -4.1 | 59 | 11 | -3.3 | 802 | 16 |
| Negative socio-emotional | 42 | 4 | 1 | 114 | 4 | 2.3 | 0 | 0 | -2.6 | 4 | 1 | -3.4 | 160 | 3 |
| Total interaction observed | 1179 | | | 3216 | | | 204 | | | 546 | | | 5145 | |

Notes
Pearson Chi-square results = $\chi^2$ = 161, 895, $df$ = 9, $\rho$ = <0.001.

Table 8.6.2 Results – rating of contractor's representatives and interaction

| Contractor rating | Most effective | | | | | | | | | Least effective | | | Total | |
| | 1 | | | 2 | | | 3 | | | 4 | | | | |
| Interaction category | No. | % | Adj. | No. | % | Adj. | No. | % | Adj. | No. | % | Adj. | No. | % |
|---|---|---|---|---|---|---|---|---|---|---|---|---|---|---|
| 1: Shows solidarity | 2 | 0 | 1.3 | 1 | 0 | -1.6 | 1 | 1 | 2.2 | 0 | 0 | -0.7 | 4 | 0 |
| 2: Shows tension release | 11 | 1 | -0.2 | 30 | 1 | -0.5 | 3 | 2 | 0.7 | 7 | 1 | 0.7 | 51 | 1 |
| 3: Agrees | 117 | 10 | 6.2 | 169 | 5 | -3.4 | 9 | 4 | -1.0 | 20 | 4 | -2.5 | 315 | 6 |
| 4: Gives suggestion | 178 | 15 | -0.6 | 535 | 17 | 2.5 | 26 | 13 | -1.2 | 66 | 12 | -2.4 | 805 | 16 |
| 5: Gives opinion | 265 | 23 | -3.1 | 858 | 27 | 1.6 | 76 | 37 | 3.8 | 135 | 25 | -0.7 | 1334 | 26 |
| 6: Gives orientation | 287 | 24 | -6.8 | 1054 | 33 | 0.5 | 78 | 38 | 1.8 | 255 | 47 | 7.5 | 1674 | 33 |
| 7: Asks for orientation | 124 | 11 | 4.4 | 239 | 7 | -0.5 | 6 | 3 | -2.5 | 20 | 4 | -3.6 | 389 | 8 |
| 8: Asks for opinion | 76 | 6 | 6.2 | 94 | 3 | -3.0 | 1 | 1 | -2.4 | 10 | 2 | -2.3 | 181 | 4 |
| 9: Asks for suggestion | 77 | 7 | 3.8 | 122 | 4 | -3.2 | 4 | 2 | -1.8 | 29 | 5 | 1.0 | 232 | 5 |
| 10: Disagrees | 27 | 2 | 0.6 | 77 | 2 | 2.2 | 0 | 0 | -2.1 | 2 | 0 | -2.9 | 106 | 2 |
| 11: Shows tension | 11 | 1 | -0.2 | 37 | 1 | 1.7 | 0 | 0 | -1.4 | 2 | 0 | -1.5 | 50 | 1 |
| 12: Shows antagonism | 4 | 0 | 3.7 | 0 | 0 | -2.6 | 0 | 0 | -0.4 | 0 | 0 | -0.7 | 4 | 0 |
| Total interaction observed | 1179 | | | 3216 | | | 204 | | | 546 | | | 5145 | |

Notes
Pearson Chi-square $\chi^2 = 253.292$, $df = 33$, $p = <0.001$.

### *Contractor's representatives' positive and negative socio-emotional interaction*

Consistent with the previous findings of the group analyses, higher levels of socio-emotional communication, when used by contractor's representatives, were associated with success. In the group analysis, higher levels of positive emotional communication were associated with successful project outcomes. The highest level of positive socio-emotional interaction was found in column one (Table 8.6.1), representing the contractors perceived to be the most effective, and those contractor's representatives that are perceived to be the least effective of this group used the lowest level of positive socio-emotional interaction. The 11 per cent of positive socio-emotional communication observed for contractor's representatives rated 1 was higher than the total percentage (8 per cent) used by the groups associated with successful outcomes (Tables 8.3.1, 8.3.2 and 8.3.3). The contractor's representatives perceived to be most effective made greater use of socio-emotional communication than the majority of other subjects observed.

The results relating to negative socio-emotional interaction are very interesting: higher levels of negative emotional communication occurred in the contractor's representatives rated more effective (representatives that are rated 1 and 2) than those rated less effective. The level of negative emotional interaction was higher than any of the negative socio-emotional interaction that was observed at the group level of analyses. The highest level observed in successful groups was 3 per cent (budget analysis). An important finding was that those representatives perceived to be less effective hardly used negative emotional categories, and also made considerably less use of the positive socio-emotional categories.

In contrast with the findings of contractor's representatives, negative socio-emotional communication was found to be lower in the groups associated with positive outcomes (two out of three analyses). To some extent the results of the group analysis were skewed by the amount of tension release, in the form of frustration, which was observed in some unsuccessful projects. Prolonged bouts of frustration, which did not result in action, did not have a positive effect on the meetings. However, the contractors' representatives rated most effective would use the negative tension to increase the importance of what was being said. When using negative emotion, the effective contractors stated the issue of concern and made it clear which party was considered responsible for the problem and the required action. The emotion was often used to emphasise the importance on resolving the matters. Emotion was evident in the tone of voice; however, the pitch was controlled and the formality and firmness with which information was presented was done in a professional manner. This was in stark contrast to the bouts of frustration observed in unsuccessful groups, which appeared erratic and uncontrolled (and one may argue unprofessional).

### *Contractor's representatives: Giving and requesting task-based interaction*

The giving of task-based communication was consistent with the behaviour of group analysis, being lower in positive outcomes (Tables 8.3.1, 8.3.2 and

8.3.3). The lowest occurrence of this category was found in the contractor's representatives perceived to be the most effective (62 per cent). The higher values occurred in the two contractor's representative groups (3 and 4) perceived to be less effective (88 and 84 per cent).

The requesting of task-based interaction was also consistent with the previous findings that were based on group behaviour. Groups that had higher levels of requesting task-based interaction were associated with positive project outcomes. Again the contractors perceived to be the most effective were those with the highest levels of requesting task-based interaction (Table 8.6.1, 24 per cent). It would seem that contractor's representatives perceived to be less effective were more reluctant to make requests to others.

The two lower rated contractor's representatives predominantly used 'giving task-based' interaction, resulting in lower usage of all other categories than that was found in the representatives that were considered to be most effective.

### Analysis of the contractor's representatives' effectiveness and interaction

A number of individual IPA categories were identified where the representatives who are historically associated with successful projects (rated 1 and 2) have higher or lower occurrences than representatives associated with less successful projects (valued at levels 3 and 4) (Table 8.6.2).

Using the Pearson Chi-square test, the results of the four ratings of contractor's representatives were found to be significantly different. The interaction profiles are shown in Figure 8.6.1.

### Category 1 (shows solidarity): Interaction of contractor's representatives

IPA category 1 does not show any noticeable trend. The results from the group analysis suggested that groups associated with positive outcomes had higher levels of IPA category 1; however, the frequency of IPA one was very low, accounting for less than 1 per cent of the total interaction observed. The results show that there were many individual professionals who did not use this IPA category. Little inference can be made on either the groups' or individual professionals' results with regard to category 1. The contractor's representatives rarely use of showing solidarity type acts (category 1).

### Category 2 (shows tension release): Interaction of contractor's representatives

IPA 2 was lower in the case of contractor's representatives rated 1 and 2 (those considered to be more effective). The results of contractor's effectiveness showed the contractor's representatives rated 1 and 2 to be lower than 3 and 4, although the percentages recorded were similar. The results from the group analysis were inconclusive regarding any trend associated with this category.

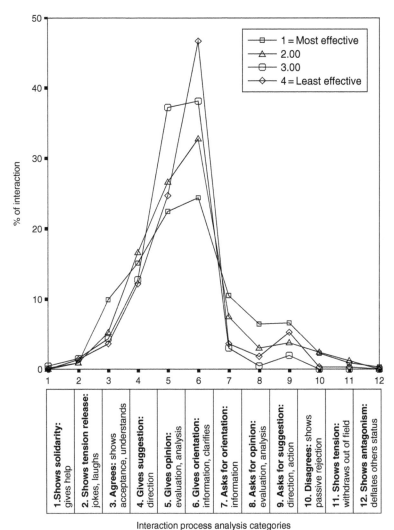

*Figure 8.6.1* Interaction profiles for the contractor's representatives' rated effectiveness

### Category 3 (agrees): Interaction of contractor's representatives

IPA 3 (agreeing) was highest in the contractor's representatives rated 1 and lowest in the contractors rated 4. Those perceived to be the most effective professionals exhibited higher levels of agreeing, the levels of interaction steadily declined as the rating of the contractor's representatives reduced (10 per cent, 5 per cent, 4 per cent and 4 per cent). This result was consistent with the trend found in the group analysis. Higher levels of interaction category 3 were associated, in two analyses, with positive project outcomes. Based on the contractor's representatives' effectiveness, there was a strong association with individuals exhibiting higher levels of category 3 and the high rating of the individuals' effectiveness.

### Category 4 (gives suggestion): Interaction of contractor's representatives

Higher levels of giving suggestions and directions were associated with contractor's representatives rated 1 and 2 (excellent = 1, satisfactory = 5). This presents a marked difference to the results found in the analysis of group inter-action. All of the statistical tests associated with category 4 (group interaction) showed significant differences between the positive and the negative outcomes. In all of the group cases IPA 4 was lower in projects with successful outcomes. However, the contractor's representatives found to be most effective used higher levels of category 4 (interpersonal interaction). If the individual contractor's representatives considered more effective gave more suggestions, yet the total number of suggestions made by the group was low, most members within the successful groups did not make many suggestions. Other members may have been inhibited or influenced in such a way that they used fewer suggestions. The most effective professionals are more likely to disagree with other proposals than other group members are – it may be that others are more reluctant to put forward proposals that are likely to be challenged.

### Category 5 (gives opinion): Interaction of contractor's representatives

No meaningful trend emerged from the category 5 results shown in Table 8.6.2. Although the values of IPA 5 used by the contractor's representatives rated 1 and 2 (when added together) were lower than the values for representatives rated 3 and 4, the individual figures vary. This was consistent with the findings of the group analysis. The total level of interaction found in the individual contractor's representative analysis was similar (26 per cent) to the total interaction found in the group analysis (27 per cent).

### Category 6 (gives information): Interaction of contractor's representatives

The levels of IPA 6 were lower in the cases of those contractor's representat-ives considered excellent (24 per cent) and gradually increased in each of the ratings, being highest in the professionals rated satisfactory (the lowest level of effectiveness) (47 per cent). The use of IPA 6 by representatives considered excellent was different from that exhibited by groups that are associated with successful outcomes. Successful M&D groups have higher levels of information giving (category 6). Thus, where contractor's representatives considered to be excellent were operating in groups associated with successful outcomes, other members in the group used higher levels of information giving. Interestingly, those contractor's representatives perceived to be most effective also requested more information (category 7). While the most effective contractor's represent-atives were not primarily responsible for the high levels of information produced in a successful project, they may have stimulated such results by being more inclined to request information than their less effective counterparts.

### Category 7 (asks for information): Interaction of contractor's representatives

IPA 7 (requesting information) was higher in the categories associated with professionals who were perceived to be most effective. This was consistent with the interaction of groups associated with projects that have positive outcomes. All of the group analyses found IPA 7 to be higher in successful projects. While the levels of information seeking were higher in groups that were successful, it was not the effective contractor's representatives who provided most of the information, although they did request it.

### Category 8 (asks for opinion): Interaction of contractor's representatives

Higher levels of IPA 8 (requesting opinions) were found in the categories associated with professionals who were perceived to be most effective. Higher levels of IPA 8 were also found in the team meetings associated with successful projects. Thus, the interaction of contractor's representatives perceived to be effective was consistent with group interaction of successful projects.

### Category 9 (asks for suggestion): Interaction of contractor's representatives

The levels of interaction associated with IPA 9 (requesting suggestions), based on the perceived effectiveness of the contractor's representatives, showed no meaningful trend (Table 8.6.2). From the group interaction data for category 9, only one of the analyses was found to be significant, with the levels of IPA 9 being higher in cases with negative outcomes. As already discussed, there was no consistent trend based on the perceived effectiveness of the representatives, although the combined value of contractor's effectiveness rated 1 and 2 are greater than 3 and 4. Contrary to the group outcome, slightly higher levels of category 9 were associated with those perceived to be most effective.

This would suggest that in all of the categories associated with requesting task-based interaction, higher levels of requesting information and suggestions were associated with the contractor's representatives perceived to be effective. The most effective contractors' representatives were good at asking questions and drawing information from the other actors. The information was then used to resolve problems and take the project forward. To ensure that appropriate action was taken, the representatives would ask those responsible for their ideas and suggestions. Rather than forcing action on others, some of the most effective representatives were particularly skilled at encouraging others to set their own course of action. Once a suggestion was made the effective contractors would develop the proposal into a firm action point – dates, times and tasks associated with the proposal would be established and recorded in the minutes.

### Category 10 (disagrees): Interaction of contractor's representatives

IPA 10 (disagreeing) was one of the most inconsistent categories associated with group interaction. Significant differences were found in two analyses, but levels were high in one positive outcome and low in another positive outcome. However, Table 8.6.2 shows higher levels of IPA 10 for contractors perceived to be most effective. Thus, disagreeing with others was found to be a trait associated with effective managers. It would seem that those less effective contractors either do not disagree with others to the same extent as the more effective contractors, or are reluctant to show when they disagree with the other professionals. Those rated more effective are more critical of others' ideas, opinions and suggestions.

### Category 11 (shows tension): Interaction of contractor's representatives

The difference found in IPA 10 was also found in IPA 11 (showing tension). In the analysis of group interaction, showing tension was significantly higher in the groups associated with projects that result in unsuccessful outcomes. However, those most likely to use IPA 11 were those perceived to be most effective. This would also suggest that those most likely to disagree were also those most likely to show negative emotions. However, as previously discussed, there are differences in the way this category is used. Some members use it to show general frustration, while others use it to express tension directly at another person.

### Category 12 (shows antagonism): Interaction of contractor's representatives

IPA 12 (showing antagonism) was observed only eight times during this research. However, it also occurred four times in the category associated with contractor's representatives perceived to be the most effective. Due to the limited data, no conclusion can be made apart from making this observation.

## Summary

The trends associated with contractor's representatives' interaction, aggregated at the negative and positive level, and giving and receiving task-based interaction level, were similar to those found in M&D group interaction. Negative emotional interactions were found to be the lowest in all situations, followed by positive emotional interaction and requesting task-based interaction, and the highest category was giving task-based interaction. The trend continues with interaction levels of most effective contractor's representatives being consistent with the group interaction associated with successful project outcomes. Positive emotional interaction and requesting task-based interaction was higher in cases that represented contractor's representatives rated most effective. Giving task-based interaction was lower in the most effective contractor's representative categories. The negative emotional category was different from the group analysis, as the group level of analysis did not provide conclusive results in this category. However, the contractor's representatives rated most effective used considerably higher levels of negative emotional communication than those rated least effective.

As with the group analysis, the extreme socio-emotional categories, 1 and 12, were hardly observed, although where they were observed they occurred more in the contractor's representatives considered to be most effective. The few observations that were found in the group level of analysis were also found in situations with positive outcomes.

Interaction categories that were found to be higher when associated with contractor's representatives rated most effective, and which followed trends of groups, included agreeing (category 3), asking for information (category 7) and asking for opinion (category 8). Giving suggestions (category 4) and showing tension (category 11) were found to be higher in the contractor's represent-ative rated most effective yet were lower in groups that were associated with successful outcomes. Disagreeing (category 10) was also found to be higher in those considered most effective, but the results of the group analyses were inconsistent. Giving information (category 6) occurred less in the most effective contractor's representatives, this being the reverse of the group findings where higher levels of category 6 were found in successful projects.

No trends were found with showing tension release (category 2), giving opinion (category 5, being consistent with the group analysis) and asking for suggestions (category 9).

# 9    Discussion of the findings

In this chapter the findings are compared and discussed against the results of earlier research. The discussion draws on the research data relating to the interaction in meetings, the differences between successful and unsuccessful projects and the interaction of the contractors' representatives.

## Participation in site-based progress meetings

### Participation and professional status

The research reviewed earlier found that group interaction is a function of those who participate within the group, and, rather surprisingly, those who do not. Before commenting on the type of interaction observed here it is worth mentioning the distribution of interaction found within the M&D team. As found in earlier group studies, participation within the M&D team is not evenly distributed and was dominated by a few professionals. Those most prominent in the group discussions included the architect, project manager and the contracts' managers. Other professionals were also proactive interactors; however, apart from those specifically identified, the other major contributors varied between case studies. The high level of architect and contractor interaction is largely due to their roles and duties. The contractor and the architect hold key positions, and much of the information, even in the relatively open context of the meeting, was directed at the architect or construction manager. Although status has previously been found to have a considerable effect on group communication, it is questionable how much the status of the professional affected participation within the M&D meeting. Due to the differences in professional standing (Higgin and Jessop 1965; Gameson 1992), it could be suggested that the architect's traditional position as team leader allows him or her greater participation rights; however, the observations are more consistent with Wallace's (1987) findings, namely that there are other control mechanisms which emerge during discussion that affect participation rights; these are discussed below. Although the interaction is skewed towards those professionals who hold central roles, it is also noted that other professionals can and do emerge as confident and influential participants. Although some professionals were reserved, others would openly contribute to discussions, and could exert considerable influence over the decisions made. Ensuring that professionals contributed to their specialist area is of considerable importance, however, it was noted that some professionals spent considerable time discussing detailed issues that were not presenting problems or making onerous contributions to issues where they did not hold relevant expertise. The failure of the chair or central professional to control and distribute participation was considered to be a contributing factor of unsuccessful projects. In two of the less successful projects, members dominated discussions even though the issues raised were not directly related to their profession. An attribute of successful groups was the ability to draw relevant professionals into discussions, ensure that due consideration was given to events and bring matters to a close. Although interaction in successful groups was distributed, the participation was not even, the professionals with the central roles were in control of the interaction, drawing others into the discussion and controlling participation, as such they were often the most dominant interactors.

### Avoiding communication

In the M&D meetings, occasionally one of the members of the group would not participate. When required to contribute to a particular point their input was often very short in duration. Some members would encourage greater participation by prompting the reluctant communicator for further information. The prompter would use various communication techniques, one of which was to ask for further information, opinions or suggestions. However, even with prompts, participation was often minimal. In the meetings that were considered successful, the other members of the group experienced greater success when encouraging members to participate. While less successful groups would leave reluctant communicators to make very sort contributions, which were of limited value, in some groups the members would persist with different tactics, encouraging the professional to make further comments. Prompts or cues that offered openings to the professional were used, questions were asked and rephrased if the response was short, once contributing supporting gestures or comments were offered to encourage further elaboration on the point.

Avoiding communication affected participation, and the individual's role within the group. Anderson *et al.*'s (1999) work on socialisation and interaction found that apprehensive people attend fewer meetings, thus reducing their potential to participate and influence the group process even further. Contrary to this view, although some of the professionals in the M&D meetings made little contribution to group discussion, their attendance at meetings was consistent with the other professionals.

There seems to be a difference between the reluctant communicator and those who choose to remain silent until they had something relevant to say or want to influence the direction of the discussion. Some members of the group exhibited periods of prolonged silence, but would be active and influential when the other members were commenting on something that could affect their interests. Then, the member would become quite vocal, being a strong participator and exerting considerable influence on discussions during this period. The members who carefully selected their timing of communication did not appear shy or uneasy with their participation, as did the reluctant communicators, but when relevant matters arose would speak confidently, using a full range of communication acts and expressions.

A few particularly strong communicators would occasionally not respond to questions raised during discussions. The short period of silence that ensued often resulted in another person raising an issue or the person who posed the question making a further comment. It would seem that the embarrassment of silence could be used to avoid addressing or responding to a particular point or cause those raising the sensitive issues to rephrase it in a way that was considered more acceptable to the non-respondent. The power of non-communication, where all communication signals are avoided, seems to result in confusion and embarrassment by the person attempting to initiate communication. Not obtaining feedback and not knowing what the other person is thinking when attempting to communicate caused the individual attempting to communicate some distress. Generally, the groups were uncomfortable with silence. The impact of the audience does seem to exacerbate the situation, and the person attempting to communicate, who does not gain a response, often rephrases the question or changes the issue being discussed. It would seem that non-communication could be an effective form of brinkmanship. Reluctant communicators acted quite differently to those who chose when to communicate or used avoidance as a negation tool. Reluctant communicators, due to their lack of effective contribution, were often guided through matters relevant to their own area of work.

### *Closure*

During prolonged periods of discussion, characterised by considerable exchanges of information, opinions, ideas and beliefs, some of the more prominent members would attempt to curtail the discussion. Those attempting to bring a closure to the problem or debate would block others' communication by disagreeing, putting forward a closing proposal or would ask others to make a firm suggestion (implying a commitment of resources) rather than allowing them to continue to offer further ideas and opinions.

Previous research on uneven participation in multidisciplinary teams has questioned the extent that problems are fully discussed and evaluated (Bell 2001). Bell claimed that when specialised members failed to participate in multidisciplinary groups the holistic perspective required could not be achieved. However, Littlepage and Silbiger (1992) found that skewed participation did not affect a group's ability to recognise individual expertise. Thus, when specialist contribution is required and other members ask and continue to ask for contributions from less dominant members, the potential to draw on the expert's perspective is increased. Also, members may attempt to close discussions when they believe that sufficient information is available to come to a decision. The results show that disagreeing, giving suggestions and asking for suggestions, which were typical communication acts used to bring discussions to a close, were more often used by those members rated most effective. The group results show that meetings on successful projects attempt to extract greater information from specific members by asking more questions. The combination of extracting information and then attempting to close the discussion may be more effective than allowing information, ideas and opinions to be openly exchanged. Higher levels of communication acts associated with giving information, ideas and opinions were found in less successful projects and contractor's representatives considered less effective. Earlier research found that individuals are more satisfied when group communication is open (e.g. Lee 1997), but the points discussed must be relevant, directed and, at some point, brought to a close. Exchanges of information need to be managed to be effective. When communication was too open and lacked direction, important matters became lost in the milieu. Obviously some matters need to be discussed in greater detail to aid understanding, but too much information, background or discussion of peripheral matters draws the direction away from the action that is required. In successful groups, simple matters were dealt with in a terse and formal fashion, whereas the few complicated matters were given more time and discussed in more detail. In groups that were effective, the group moved quickly through the matters considered simple or easy to resolve and focused on those requiring attention, but were quite firm on bringing matters to a positive resolution.

Those rated the most effective were often influential in the group, and although they did not block participation, they ensured that any matters that distracted the conversation were quickly curtailed. The control of communication was often used to include members who were not strong communicators, but possessed expert knowledge. Participation in groups that controlled the discussion was often more distributed across all members, whereas, when communication was open, free and uncontrolled reluctant communicators would often be excluded from the conversations with the majority of the interaction being controlled by dominant members.

### *Positive and negative socio-emotional communication norms*

Previous research found that the use of socio-emotional interaction within groups is considered to be essential for building and repairing relationships (Bales 1950, 1953; Poole 1999). A number of socio-emotional interaction patterns were found that were considered to be characteristic of the M&D team meeting.

Trends in emotional interaction were particularly strong. Positive emotional communications were consistently higher than negative socio-emotional acts in all of the meetings observed. However, the total amount of emotional communication was low, being considerably lower than task-based interaction. The general low level of socio-emotional interaction is a very important finding. Earlier research identified a number of reasons for low levels of socio-emotional communication occurring at different stages in the group's processes. Interaction patterns are distinguished by high levels of task-based interaction and low levels of socio-emotional interaction during the early stages of group development (Heinicke and Bales 1953). The transcripts and audio recordings used by Gameson (1992) may have limited his observations of socio-emotional behaviour during potential clients' and construction professionals' first meetings, although the very low levels of socio-emotional interaction recorded may be due to the lack of group development. Lower levels of socio-emotional interaction are associated with newly formed groups (Bales 1950; Hoffman and Arsenian 1965; Schutz 1973; Wallace 1987).

Bales (1950, 1953) claimed that in newly formed groups, or groups that experience personnel changes, members avoid using emotional interaction until they are aware of how other members will react. As groups develop and become aware of the group's regulatory framework, the amount of socio-emotional communication increases (Heinicke and Bales 1953); however, if groups change they tend to regress back to early stages of group development (Borgatta and Bales 1953). Wallace (1987) found that when a group changes, or a member leaves or enters a group, the group behaves as if it has not previously worked together before. When groups regress back to the early group development phases, interaction is categorised by high levels of task-based interaction and low levels of socio-emotional interaction (Bales 1950, 1953; Wallace 1987). The findings of this research support previous studies and would suggest that the unstable, short-term nature of construction teams is not enabling the group to develop high levels of socio-emotional interaction. Other research on temporary multidisciplinary work groups has found similar interaction patterns, with discussions being predominantly task-based: group members concentrating on issues, facts and information rather than showing emotion (Bell 2001).

The low levels of socio-emotional interaction found in this study are typical of this environment (occurring in all meetings). It is suggested that the low emotional interaction is due, at least in part, to group changes and/or the temporary nature of the team (not having been together sufficiently long enough to develop their interaction fully). However, when examining the series of three meetings in each case study, the level of socio-emotional communication did not increase from one meeting to the next. The levels of emotional communication were sporadic, with higher and lower levels of emotional communication occurring in various meetings. Even when the emotional exchanges were high they were relatively low compared with most studies of adult groups.

A detailed investigation of emotional communication in the M&D meetings revealed that the positive socio-emotional communication exceeded that of the negative socio-emotional communication threefold. Positive reinforcement (using positive emotional interaction) such as agreeing, showing solidarity and being friendly were used to maintain relationships.

Most of the construction professionals restrict their use of emotional exchanges. As the group members have little experience of other members, there are reasons to suspect that they are uncertain what effect the use of emotional interaction would have on the professional relationship. Previous research has suggested that some members may take advantage of this position, being more proactive communicators (Reichers 1987; Anderson *et al.* 1999). By participating more in the group's interaction, individuals develop a greater awareness of the group dynamics and other group members' behaviour towards themselves. As individual members become comfortable with interacting, and more aware of the group's regulatory framework, they engage more assertively, affecting the development of relationships and group goals. Those who are reluctant to use emotional exchanges do not know or learn what effects emotional communication acts have on others and themselves. The members rated more effective used a higher level of socio-emotional communication and were influential in the inclusion of other members. By helping others participate and bring matters to a close the members rated most effective had some control over the development of the group's regulatory framework.

### Task-based communication norms

In all of the M&D meeting interaction profiles, the levels of interaction associated with giving and requesting task-based interaction were very high. The levels of giving information (category 6) were higher than all of the studies previously conducted by Bales (1950, 1970). Given that previous research has found that decentralised communication structures, such as meetings, are considered to be most effective when dealing with complicated issues and high levels of information, the M&D meeting provides an effective forum to deal with this situation.

Because there are different specialists in the M&D meeting, many of the issues raised by a specialist needed further elaboration to ensure that all parties, who have other specialisms, understand the terms used and the issues being discussed. The levels of opinions, explanations and analyses were also high; this type of behaviour would typically be used to develop arguments or support the understanding of information. Much of the additional explanation, used by a specialist, emerged as messages were received, indicating that initial information was unclear to the other professionals. As participants present information to the group, other members observe the facial expressions and body language of others, helping to indicate the other parties' levels of understanding or not; thus, explanation of an issue may continue until signals received from listeners confirm understanding.

The need to build on information to ensure that a relevant context is created was identified in the literature on human communication (Sperber and Wilson 1986). New assumptions can only be made once a contextual understanding is developed. Occasionally, more information than is provided by an initial explanation is needed in order to understand a problem or situation. At this point, participants receiving information may face a dilemma, as Lee (1997) found, when professionals do not understand, they may be reluctant to ask for help or further information in order to understand better.

The levels of question asking observed in this professional environment were low considering the high levels of information, and this is consistent with research on question asking in other organisational contexts (Capers and Lipton 1993; Lee 1997). Although low, it is clear from the M&D group profile of interaction that members of the group occasionally request further information, opinions and suggestions. Asking for suggestions is much higher than that found in work by Bales (1950, 1970) and Gameson (1992).

The M&D meetings were often used to resolve problems. Problem-solving often involves change and, as previously suggested, professionals may resist change as it leads to a redistribution of resources (Loosemore 1996a). The requesting of information and opinions was often used to draw further information and explanation on a specific situation into the forum. However, the requesting of suggestions was slightly different, often being used to make people commit to a particular argument or direction. This communication tactic was frequently used to place pressure on other people to make a decision, suggest action and encourage others to commit their resources.

### *Relationship between task and socio-emotional interaction*

Earlier research indicated that the discussion of technical problems results in the emergence of socio-emotional reactions. As people discuss task-related issues, tension is built up which is then released through positive and negative socio-emotional interaction, positive emotional interaction being used to disperse tension. The levels of emotional interaction observed were low when compared with earlier findings, and extreme expressions such as showing antagonism or solidarity were hardly observed. However, it was noted that even one or two emotional exchanges had a real and very noticeable impact on the group's behaviour. For example, when a professional expressed tension, anger or frustration (recognised by vocabulary, tone of voice or facial expression), all members of the group would become much more attentive. Group members not involved in the emotional exchange adopted motionless postures as they listened carefully to the interaction. Background noise, such as rustling of papers and informal comments between adjacent professionals, did not occur during emotional encounters.

With the vast amount of information dealt with during M&D meetings, and with the adversarial nature of construction, one would expect to find considerable levels of conflict, yet conflict in the M&D meeting environment is low. In fact, the amount of conflict, or negative socio-emotional communication, was lower than reported in previous studies of adult groups (Bales 1950, 1970; Gameson 1992). However, the first case study findings supported Lawson's (1970, 1972) observation that increases in information results in increased levels of conflict. As information presented in the M&D meeting reduced, and fewer persons were present in the meeting, the amount of disagreement and the levels of conflict also reduced.

Research by Pavitt (1999) has found that for group satisfaction and successful group outcomes there should be several times more positive than negative emotional interaction. Furthermore, task-based interaction should be twice that of emotional interaction. High levels of positive socio-emotional interaction are claimed to be necessary to maintain relationships (Bales 1953; Pavitt 1999);

however, overuse of agreement can result in 'groupthink' (Cline 1994). In the M&D meeting, while the positive emotional levels were three times higher than the negative emotional levels, they were not as high as previous findings would suggest are necessary for group satisfaction; however, the level of criticism found would help to avoid groupthink. As the amount of negative emotional communication increased, the level of positive socio-emotional communication also increased in many of the meetings. Thus, the M&D participants seem to focus on ensuring that relationships are repaired following conflict. The relationship between successful outcomes and positive emotional interaction is discussed in detail later.

## Differences between interaction in successful and unsuccessful projects

### Individual socio-emotional interaction

The individual emotional categories showing solidarity, agreeing and showing antagonism are consistently and significantly higher in the positive project outcomes, and showing tension is consistently and significantly higher in all of the negative project outcomes. The amount of positive socio-emotional interaction is significantly greater in successful than unsuccessful projects.

A trend also exists for greater use of a broader range of socio-emotional categories in successful projects. Unsuccessful project groups tend to limit socio-emotional communication to the more neutral acts (disagreeing and agreeing) and restrict their use of more extreme categories. It is important to note that, even though the successful projects used less communication acts (Table 8.4.2), the group communication was distributed across more categories, with a few occurrences in the extreme emotional categories 1 and 12. Successful projects make greater use of the more neutral and extreme emotional types of interaction.

### Positive and negative socio-emotional group interaction

In all of the analyses of positive and negative emotional interaction, the amount of positive socio-emotional interaction is significantly higher in the cases with positive outcomes. The results for the negative socio-emotional communication are not as consistent as for the other categories. In two analyses, duration and conflict with the architect, the amount of negative socio-emotional communication is lower in projects with successful outcomes, but in the budget analysis, the negative socio-emotional interaction is higher in successful outcomes.

Case study 3 has the second highest occurrence of disagreement. Factors that contributed to this included the amount of information that was unsupported by explanation, and the adversarial relationship occurring between the client's project manager and the architect. Compared with other studies, the amount of socio-emotional interaction observed in Case study 3 was high. Much of the emotional expression was due to tension experienced between the architect and the project manager and others using positive socio-emotional categories to ease the tension. Even though the socio-emotional interaction is high, the positive interaction is much higher than the negative emotional interaction. As positive emotional interaction is often used to build and maintain relationships (Bales 1950, 1953; Ellis and Fisher 1994), the high levels of positive emotional interaction may have helped the group recover from the negative emotional outbursts. Even with the high level of negative emotional interaction the project was completed on time, within budget and without experiencing contractual or legal conflict with the client or the architect.

From the qualitative observations, negative emotions were used in two ways: to express frustration, but without showing who or what is causing the frustration, or to express disagreement, concern or even anger, which was directed at others because of what they have said, done or conversely not done. When a level of negative emotion was directed at a particular group or individual, it would normally result in a response related to the issue of concern. Actors that expressed their frustration, but did not direct the negative emotion at a group or individual, rarely received responses relevant to the issues raised. When frustration was used without a specific focus or direction, members would allow the frustration to be vented, but did not ask the person to clarify their concerns or offer assistance. Professionals from other organisations involved in the project rarely volunteered to help those in distress. Case study 7 was an example of such behaviour. The contractor's project manager expressed considerable frustration with the project; however, it was not clear who or what the frustration was aimed at. Case study 7 did not complete on time, or within budget, and it did experience conflict with the architect and the client.

In other cases, when frustration, anger or disagreement was vented in response to another person or a particular situation, those responsible were immediately prompted into defending their actions or statements. As arguments developed, professionals would often alter their position, taking account of the dissent shown. From qualitative observations, negative emotional expression had a greater effect on the group interaction when it was directed at an individual, or organisation. However, the need to repair relationships after such encounters was also noted. Following disagreements, positive socio-emotional interaction was often used to disperse the tension.

In Case study 1 a trend was identified between the amount of problems faced by the group and levels of negative emotional interaction. As the number of issues reduced, the level of negative emotional interaction reduced. Indeed, the problems faced by the group in Case study 1 reduced to such an extent that no more scheduled M&D meetings were necessary. Successfully overcoming problems meant that professionals had a tendency to agree on future direction. The number of people present at this meeting also reduced as problems reduced. Reduction in problems and the amount of information presented in a meeting appears to reduce the potential for disagreement.

Other reports, previously discussed, suggest that groups associated with desirable outcomes have considerably more positive (categories 1, 2 and 3) than negative socio-emotional communication (categories 10, 11 and 12) – positive socio-emotional interaction being used to repair relationships. The difference between positive and negative emotional interaction (Tables 8.4.1, 8.4.2 and 8.4.3) for the M&D meeting context is much less than has been experienced in other adult groups. The differences associated with positive and negative outcomes in the M&D meetings tend to be only marginally different in percentage terms, although statistically significant. The interaction norms of the group are very strong and small changes in the behaviour of the group are significant. Although quantitatively differences were small, qualitative observations suggest that any negative or positive emotional exchange had a notable effect on the group.

Analysis of the profiles for each individual meeting shows that greater amounts of positive socio-emotional interaction compared with negative socio-emotional interaction has a stronger association with successful groups. The interaction data presented in Table 8.3.1, from the projects completed within budget, exhibit a small but significantly higher positive socio-emotional communication (showing solidarity, tension release and agreeing) than the projects that were over budget.

The levels of conflict, disagreement and tension shown in successful and unsuccessful projects vary. In two of the analyses (scheduled duration and conflict), the levels of negative emotional interaction were significantly lower in the successful outcome and in one analysis levels were higher (budget). Although disagreements are more prominent in successful projects, conflict occurred in most projects. However, in the more successful projects when discussions emerged which involved disagreements, there was a greater tendency to recover from the argument with an agreement, making a light hearted statement, joking or other humorous acts. When the level of negative interaction increased in the successful projects, the positive emotional interaction in these groups increased considerably more. In successful groups when negative interaction increases, as it did in the budget analysis, the group interaction compensates for this by much greater increases in the positive emotional interaction. Thus, the difference between positive and negative interaction is maintained ensuring that relationships are sustained.

In Case studies 2, 4, 6 and 8, it was noted that interaction associated with the more extreme socio-emotional categories, such as shows solidarity, raises others' status, gives help and shows antagonism, deflates others' status, were not observed. These projects tended to be unsuccessful. In Cases studies 4 and 6, the construction projects were not completed within budget or during the scheduled time, and in Case study 8 the project did not complete within the scheduled duration. The more extreme categories of emotional interaction tended to occur more in projects associated with successful outcomes. Contrary to many government reports and common belief, much of the conflict observed was useful in exposing problems and initiating action to resolve problems and differences. As Loosemore *et al.* (2000) claims, conflict, if handled correctly, has benefits. Negative emotions need to be managed and used to benefit the project, rather than avoided. Clearly conflict that develops into a formal dispute is not conducive; however, attempting to remove all forms of conflict may be detrimental to the construction industry. There are subtle differences in the ways groups engage in and manage conflict. Successful groups engage in conflict and ensure that relationships are maintained and supported.

## Summary: Task-based interaction categories

The differences found in the use of statements that request and give information between successful and unsuccessful projects were significant and consistent. Higher levels of requesting task-based interaction and lower levels of giving task-based communication acts were associated with successful outcomes. Also, the total amount of task-based communication (including both giving and receiving type acts) was lower in projects with successful outcomes.

Successful projects are associated with comparably lower levels of giving task-based acts than used in unsuccessful projects. However, on closer examination of the research data, giving information and opinions is significantly higher in the successful cases, but giving suggestions (category 4) is consistently and significantly lower. Giving suggestions has high levels of adjusted residuals showing considerable difference between the successful and the unsuccessful projects. This would indicate that much of the difference associated with giving task-based interaction was attributable to giving suggestions. Successful project teams have a strong tendency to give fewer suggestions than those associated with unsuccessful projects, this is compensated with slightly higher levels of information and opinion giving. Initial observations would suggest that this contradicts previous group studies which have found that when group members provide greater direction and control their group's performance is improved. However, individual members within successful groups do offer high levels of direction (see discussion on contractor's representatives). Control and direction of successful groups was achieved through a more distributed use of the task-based interaction categories. Requests for information, opinions and suggestions as well as giving suggestions were used to involve relevant participants in the formation of action points. Each of the task-based interaction categories was used to extract information and provide direction. Although project managers would give information so that project parameters were established, those responsible for an area of work would be asked to provide suggestions of action rather than being provided with a set course of action by the project manager. In less successful projects, the project managers or influential members would attempt to impose deadlines and action points on others. It was common to find that in subsequent meetings the deadline set had not been achieved. When the participant worked to establish compatible action points, with the individuals responsible for the action being asked to provide the date for action, it was more likely that the date would be achieved. It should be noted that in successful projects dates for action were always established, recorded in the minutes and tracked.

Case study 3 has the second highest levels of giving information (category 6). Giving information is considerably higher than giving opinion, evaluation and analysis (category 5). Although this is found in other cases observed, it is not typical of many previous studies. In a previous study, where the levels of information were found to be greater than the levels of explanation, it was suggested that this situation was possible due to the high level of understanding shared between the participants (Bales 1950). The level of explanation offered by professionals in the M&D context is relatively low. A considerable amount of terminology is often used which may not be understood by all of the professionals. However, while not overtly asking others to explain technical terms, members in successful groups were more likely to ask for greater information and explanation when discussing issues.

Giving opinions and explanations (category 5) occurred the most in Case study 2. This figure is possibly skewed by interaction during meeting 2.1 where observations of category 5 accounted for 47 per cent of the communication acts. The amount of analysis, evaluation and opinion during this meeting was high compared with other meetings; only Case study 9 has a similar level of interaction in this category. Interestingly, both projects were completed within budget but not within the scheduled duration. Bales (1950) noted that the act of giving opinions often results in or encourages others to give opinions, and this is particularly true for Case studies 2 and 6. In summary, high levels of interaction associated with opinion and information giving, and lower levels of giving suggestions, were more common in the projects with successful project outcomes.

Categories 9 (asking for suggestions), 8 (asking for analysis) and 5 (giving analysis) are not consistent across all of the results. However, as previously mentioned, the overall level of question asking is significantly higher in successful projects.

## Interaction of contractor's representatives

The professionals rated as most effective are those who have contributed to or led successful projects. Their actions and behaviours are considered to have made a greater contribution to the success of the projects that they have previously been involved with than those rated lower. In many respects, these professionals can be considered to be informal or elected leaders within the project domain. Their interaction practices will form one of the traits that contribute to their performance. The analyses have found significant differences between the interaction of those considered most effective and those who are less effective. Four of the professionals were rated 1 (most effective), nine were rated 2, eleven were rated 3 and two were rated 4 (least effective).

Some of the interaction characteristics of those individual contractors considered most effective are different from the interaction patterns of the successful groups, even though many of the effective contractors were members of the project teams that successfully completed projects within time and budget.

The differences between the individual and group interaction may be due to the perceived status of an individual (Collaros and Anderson 1969; Hare 1976; Brown 2000). However, consistent with Wallace's (1987) observations, rather than perceptions of status that may exist before the project commences, it is believed that the assumed status, role and authority of contractor's representatives

emerge during group interaction. By interacting with others, contractor's representatives attempt to co-ordinate and control project activities. Contractors use
their communication to develop their role, exert their authority and attempt to
influence the management and design team so that the group's actions result
in performance that is in the contractor's interest. Thus, those considered more
effective may adopt different interaction styles from the less effective contractor's
representatives, enabling them to produce more favourable results.

Group norms affect all members of the group (Keyton 1999), yet high status
members may be exempt from the norm (Hackman 1992). Such deviation can,
if accepted, influence others, changing the group norms and the decision-making
process (Janis 1982; Senge 1990; Hackman 1992). Thus, those who interact
differently from the group norm may have considerable influence on group
behaviour. The interaction patterns of those considered to be most effective are
extremely interesting as they offer characteristics of positive interaction behaviour
(from the contractor's perspective).

### *Contractor's representatives' participation*

The participation of contractor's representatives differed during meetings
depending on their rating. The average number of communication acts made
during three meetings by the most effective contractors is 295, the next highest
rating is 292, those rated 3 contributed the least, with 23 communication acts
over three meetings and those rated 4 used an average of 273 communication acts
per representative. From a descriptive perspective, while there is little difference
between professionals rated 1 and 2, and a small difference between two and
four, those rated 3 seem reluctant to participate or did not have anything useful to
contribute in meetings. The relationship between participation and effectiveness
is not clear. Although Bales (1953) and Hare (1976) suggests that those who talk
the most become leaders and have a greater influence over the decision-making
process, in this study those perceived to be least effective also have high levels
of participation rates.

Those rated three and four do not disagree or engage in the same amount
of conflict as the higher rated contractors. This could be due to contractor's
representatives not having alternative arguments, or being reluctant to disagree
so that they avoid conflict. Earlier research (Cline 1994) found that not fully
evaluating proposals could result in unsuccessful outcomes. Failing to challenge
others or disagree has been previously found in group members associated with
reluctant communicators (McCroskey and Richmond 1990; Haslett and Ruebush
1999), although only those rated 3 could be classed as reluctant communicators.

### *Effective and less effective contractor's representatives: Socio-emotional differences*

As previously discussed, slight changes in positive and negative emotional interaction can have a considerable impact on subsequent reactions of the group (Hare

1976). Those professionals who have performed better in projects use higher levels of emotional interaction. However, the high level of emotional behaviour used by the most effective professionals was not limited to positive emotion, they also used significantly greater amounts of negative emotion. For the professionals rated one and two, negative emotional interaction accounted for 4 per cent of their interaction, while those rated three and four used less than 1 per cent of their total interaction for negative emotion. A closer examination of the individual categories also reveals that those considered most effective use higher levels of the extreme positive and negative categories, as well as those emotional categories that are considered more neutral. However, the occurrence of these categories is low in comparison with task-based categories. Those considered more effective engage in moderate levels of conflict.

When extreme emotional behaviour (categories 1, 2, 11 and 12) was observed in meetings, other professionals became quiet and focused on the communicator. In much the same way that Sperber and Wilson (1986) discussed the nature of highly salient events, the emotional behaviour of professionals draws the attention of other professionals; it arouses interest causing those who can hear the interaction to focus, observe and listen to what is being said and communicated. Towards the end of meetings professionals became less attentive – participants would look out of windows, gaze into space, look at their watches and look at documentation not connected with the meeting. However, when emotional exchanges were made, especially extreme behaviours, all of the group members would focus on the interaction and the professionals involved. Those to whom the emotional interaction was aimed became much more attentive. Negative emotional interaction that was specifically directed at a person suggests that the initiator is concerned with the current or past behaviour of the receiver. It provides a potential threat to the social and business relationship, and may question technical expertise. Receivers of negative emotional interaction react by defending their behaviour, counter attack with further negative emotional interaction or may suggest action to remedy the situation. Thus, group members are more likely to entertain (pay attention and listen) emotional encounters; such exchanges were more salient than other communication events or discussions that may pass as background information or even noise. Those considered more effective use greater levels of emotions. Considering the point made above, when using emotional interaction other group members react. Thus, when those considered more effective want others to listen to what is being said they are more inclined to use emotional tones to ensure they get a reaction.

The negative and positive emotional exchanges often formed part of detailed discussions and arguments. Barge and Keyton (1994) suggest that people use argument to get actions to be viewed in a particular way. The contractor's representatives used negative emotional behaviour to show how serious or important the statement was and to make sure others listen to what they are saying. Those rated more effective were more inclined to engage in such acts. However, negative emotions may threaten relationships or strengthen them. While Wallace (1987) and Cline (1994) claim that negative emotions may be avoided to maintain relationships, Averill (1993) suggests that when negative emotions are expressed, in view of a perceived wrong and not to attack the other person, relationships are often strengthened more often than they are weakened. Emotional interaction provides signals of how satisfied group members are with the behaviour of others. Early expression of negative emotion allows others to realise the dissatisfaction and, if necessary, enabling them to alter behaviour and repair or maintain the relationship. The effective contractors were more likely to show disagreement than those rated less effective.

The contractors perceived to be the most effective used the most positive emotional expressions. As already discussed positive socio-emotional behaviour is used to build and maintain relationships, and disperse tension (Bales 1950, 1953, 1970; Hare 1976; Frey 1999; Keyton 1999; Mabry 1999; Poole 1999; Schultz 1999). Although the contractors are likely to use greater levels of negative emotional interaction, thus creating tension, they are also more likely to use increased levels of positive emotional exchange to dissolve negative emotions and rebuild relationships. Thus, the emotional exchanges of effective contractors are changeable, responding to the nature of the situation and information received. This supports Gibb's (1961) theory that good communicators develop intuitive and flexible interaction techniques that are responsive to others.

If professionals do not break or resolve negative emotional interaction, the communication environment may become very defensive. As Gibb (1961) suggests, if negative and defensive actions take place without question, an increasingly circular destructive response may occur. Effective contractors use negative emotion; however, they use greater amounts of positive emotion to ensure that the conflict does not threaten the relationship.

### *Effective and less effective contractor's representatives: Task-based differences*

The contractors rated most effective provide less task-based interaction but request greater levels of task-based interaction than those considered less effective. A more detailed analysis of the individual categories reveals that this is particularly true for category 6 (giving information) with those considered least effective providing twice as much information as those considered most effective, but this is not true for giving suggestions. The level of suggestion and direction giving is greatest in those rated one and two. The directional difference of category 5 (giving opinions) is inconsistent.

Requesting information, opinions and suggestions are highest in those rated most effective. Such behaviour is obviously useful for drawing on the knowledge of group members; as Ellis and Fisher (1994) suggests, asking questions is the single most effective way to extract ideas and information from other members of the group.

The contractor's representatives considered most effective attempt to gather information from the group: they request very high levels of information and greater amounts of opinions and suggestions than do less effective representatives.

The level of information and opinion giving is quite low, however the percentage of suggestions used is high; the professionals considered most effective request high levels of information, opinions and suggestions, but also offer direction to the group.

When effective actors are present in groups that produce successful outcomes, individually they offer considerable direction, yet the total suggestions by the group tends to be lower in proportion to other communication acts used compared to groups that are associated with negative outcomes. In unsuccessful project teams, too many members make suggestions. Contractor's representatives rated more effective attempt to control contributions of other members by asking for information, opinions, suggestions, and disagreeing and agreeing. Although more effective members do ask for suggestions, they use higher levels of asking for information. Such behaviour helps control the amount of suggestions presented by others. In less successful groups, suggestions may be offered prematurely without considering relevant information.

Hare (1976: 78), comparing the research of Parsons, Couch, Bales and Bion, proposed that giving suggestions is a dominant behaviour by a group member who is attempting to make a decision. Informal and formal leaders have been found to initiate suggestions (Heinicke and Bales 1953). Hare's (1976) review of research on influence and power noted that when a person repeatedly attempts to influence another they are more likely to be successful. The repeated use of suggestions by those M&D professionals considered most effective increases the potential of the suggestion being accepted.

The analysis also suggests that those considered most effective encourage and prompt greater information exchange and play a considerable part in offering direction to the M&D team. Shepherd (1964) highlighted the importance of ensuring that relevant information is not withheld, and Hare (1976) also claimed that the direction and structure of the group is also an important character-istic of success. The interaction behaviour of those considered more effective draws information from the M&D team and attempts to direct the group towards a suggested course of action. Those considered most effective make greater attempts to control group interaction by using a broad range of communication techniques.

# 10 Conclusions and recommendations

Many construction sector reports have claimed that the interaction between management and design teams are often conducted in an adversarial environment, with interaction being characterised by tension, negative emotion and conflict. Contrary to such perceptions, the data collected from the progress meetings did not reveal high levels of emotional interaction, but more subtle traits of interaction. Although the research findings are based on a small sample of meetings compared with the number of meetings that take place on a daily basis, the findings tend to suggest that the behaviour of the participants, and thus the culture that exists within, and between, progress meetings may be less adversarial than the reports allude to. Indeed, the research shows that there is a lack of emotional expression within meetings.

The groups and individuals that performed best were found to use a wider range of communication acts. In some cases, instances of what has been described as uncontrolled negative emotion, in the form of frustration, were expressed, but it had little impact on the other members of the team. The best performers were efficient in their use of discourse, being direct, seeking relevant information and suggesting or requesting action in a timely manner. The effective groups used less interaction than the groups that did not achieve satisfactory performance. Groups that were effective used a broader range of communication techniques. When the most effective individuals expressed emotion, their comments were focused and controlled, whereas the negative emotional communication used by less effective members lacked direction, was often prolonged and confused. Groups and individuals considered more effective use a broader range of communication techniques and have to use less communication acts to achieve their desired outcome.

Our review of earlier research found that meetings and face-to-face communication were an important part of human communication in construction teams. This was supported in the research reported here. Interaction during construction project meetings was used to resolve technical problems through information sharing and discussion, that is communication was used to facilitate decision-making. Interaction was also used to assist in the development of professional relationships, thus helping to deliver projects in accordance with planned objectives. Some subtle, yet significant, differences were identified in the communication traits of the participants involved in successful and less successful projects. These subtle differences were also found to exist between the effective and the less effective contractor's representatives. The results help to show how different groups work through, and respond to, issues and problems raised during meetings.

## Typical management and design team meeting interaction

The first aim of this research was to determine whether the patterns of group interaction observed were specific to context in which they occurred. The research objective was to determine whether *the management and design meeting has a characteristic model of interaction, being different from interaction profiles found in other contexts.*

The M&D team meetings did have a number of characteristics that set it aside from previous observations of other types of group interaction. The patterns that are most consistent (typical) across all meetings are considered to be the interaction norms associated with this group context. The findings presented here are consistent with previous research on interaction norms. They support the notion that individual members of the M&D team are aware, either sub-consciously or consciously, of interaction norms and accommodate behaviour that conforms to these norms. These behaviour and interaction norms provide a type of regulatory framework in which communication takes place.

### *Task and socio-emotional interaction norms*

Group interaction is characterised by high levels of task-based interaction with low levels of emotional interaction, which was consistent across all observations. Levels of task-based interaction are typically higher than the socio-emotional communication in adult groups, although the levels of socio-emotional commu-nication observed were found to be lower than those previously observed (with the exception of Gameson's sample).

### *Inferences made regarding the nature of interaction norms*

In all of the interaction models, the levels of interaction associated with giving and requesting task-based interaction were very high. The levels of giving information were higher than in many of the earlier studies; however, the few studies of real 'work-based' meetings do have similarly high levels of task-based interaction. The lack of research undertaken in other commercial environments makes it difficult to state whether the concentration on task-based interaction is peculiar to the M&D meeting.

Because there are different specialists present in the M&D meetings, many of the issues raised by a specialist needed further elaboration to ensure that all parties understood the terms used and the issues being discussed. The levels of opinions, explanations and analysis were high: this type of behaviour would typically be used to develop arguments or support the understanding of information. Much of the additional explanation used by a specialist emerged as messages were received, indicating that initial information was unclear to the other professionals. Thus, complex discussion between different specialists requires prolonged explanations that are supported by large amounts of information.

The progress meeting was often used to resolve problems. Problem-solving often involves change and, as previously suggested, professionals may resist change when it leads to redistribution of resources. Requesting information and opinion was often used to bring further information and explanation on a specific situation into the forum. However, the requesting of suggestions was slightly different, mainly being used to make people commit to a particular argument or direction. This communication tactic was frequently used to place pressure on others to make a decision and/or suggest action and commit resources.

Earlier research found that interaction related to asking for help and asking questions is uncommon in professional environments. This behaviour was consistent across all of the meetings. Overtly asking for help was not observed in any meetings. Also, the levels of requesting information, opinions and suggestions, which presented a way of seeking additional information, were considerably lower than communication acts that provided information.

Laboratory research and studies of counselling sessions found that the amount of positive emotional communication should be several times greater than negative emotional interaction if the group is to operate effectively and maintain relationships; however, such high levels of positive emotion do not occur in work groups. The results obtained from the M&D meetings are more comparable to other studies of work groups. While higher levels of positive socio-emotional interaction are associated with successful projects, the differences between positive and negative interaction are relatively small. However, the use of positive and negative emotional interaction in the M&D (commercial) environment had considerable impact on group interaction. Participants who used direct and focused emotion to express themselves achieved greater cooperation and responsiveness from other members of the team compared with those who did not use emotion, or failed to control their use of emotion.

The general low level of socio-emotional interaction is important. The findings of this research support previous studies and suggest that the unstable, changeable, short-term nature of construction teams is not enabling the group to develop high levels of socio-emotional interaction.

This research found that group interaction in the M&D team meeting has a 'typical' (characteristic) model of communication. At the aggregated levels of socio-emotional and task-based interaction the categories are consistent across all observations, and some of the individual IPA categories also have high levels of consistency across all meetings. The first aim of the research was to determine whether *the management and design meeting has a characteristic model of interaction, being different from interaction profiles found in other contexts.* The research shows that interaction across the meetings has consistent patterns of behaviour that are different from those found in other contexts previously studied.

While the aggregated levels of task-based and socio-emotional interaction remain consistent across the meetings, it is noted that the levels of socio-emotional behaviour are much lower than previous studies. It is noted that this research focused on *bona fide* meetings that had business and commercial pressures and most of the previous communication research has been conducted in laboratories or non-business environments. It is believed that interaction in professional work-based environments will tend to exhibit less socio-emotional interaction. Also, the temporary nature of many project groups may suppress the development of socio-emotional interaction.

## Conclusion: Hypothesis 1

The following section identifies the group interaction variables that were found to be significantly different when they were attributed to successful or unsuccessful project outcomes. Hypothesis 1 stated that *the interaction patterns of group meetings associated with successful project outcomes are significantly different from the interaction patterns of group meetings that are associated with unsuccessful project outcomes.*

The hypothesis is supported by the research findings. Significant differences were found between the interaction associated with successful and unsuccessful projects. A significant difference was found when examining the positive and the negative socio-emotional elements, and the giving and requesting task-based interaction. Significant differences were also found between several of the individual IPA categories. Furthermore, several important observations were made that support the hypothesis.

The results support earlier research. Higher levels of positive emotional communication tend to be associated with positive group outcomes, helping to maintain relationships and recover from conflict. Although significant differences were found between the amounts of negative emotional interaction in different analyses, the direction of the differences varied between cases and in some analyses the differences were so small that no real inference could be drawn. The directional differences of negative emotion between successful and unsuccessful projects are of little practical use. The important finding regarding negative socio-emotional interaction (such as disagreements, tension and conflict) is that it occurs in both cases of successful and unsuccessful outcomes; however, in the successful outcomes the amount of positive emotional interaction is much greater, thus helping group members to recover from the conflict. Disagreeing with others and engaging in conflict were useful in exposing differences, identifying problems and has the potential to reduce groupthink. Uncontrolled negative emotional expression has little useful impact on a group; however, focused and controlled negative emotion tends to result in participants responding to the issue raised and action being proposed. In all cases, when negative emotion was focused participants defended their position, but through the conflict and recovery an amicable resolution was proposed. The successful groups used positive emotion to reduce tension after negative episodes of emotion. Expressing concern, worry, fear or showing anxiety without making it clear who or what was the cause of the problem was not an effective method of problem resolution. Expressing too many issues at one time also reduced the ability of the group to resolve the problems.

### *Model of group interaction*

The differences between successful and unsuccessful group interaction is modelled in Figure 10.1. The level of positive emotion (agreeing and being supportive) is greater in successful M&D teams. The level of negative emotion (disagreement, showing tension and conflict) tends to be slightly lower in successful project teams, although it does exist and, more importantly, it is essential that people show when they do not agree with an idea, proposal or argument. Conflict helps to evaluate suggestions and expose weaknesses. The use of positive emotion helps to build and maintain relationships between the members engaged in what may be stressful discussions. Successful teams use

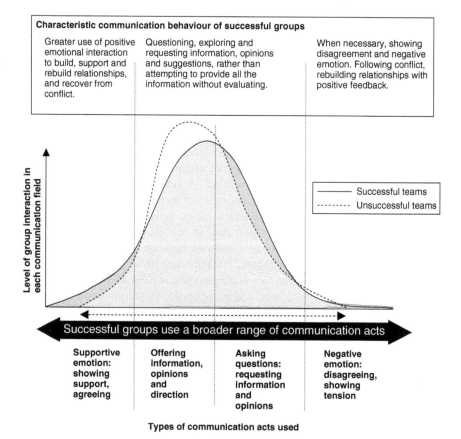

*Figure 10.1* Successful and unsuccessful management and design group interaction

more emotional interaction, occasionally showing extreme emotional expression, such as showing solidarity, being friendly and showing anger and tension.

Unsuccessful groups use less emotional interaction and more neutral emotional expression, limiting such interaction to agreeing or disagreeing. Although successful teams use slightly less task-based discussion, they do ask more questions. Question asking encourages more people to share their ideas and suggestions. Unsuccessful groups tend to concentrate on task-based interaction, with high levels of information, ideas, suggestions and opinions, low levels of question asking and little emotional expression. The limited emotional interaction reduces opportunities to build and maintain relationships, and manage conflict.

## Conclusion: Hypothesis 2

The final hypothesis stated that *the contractor's representatives considered to be the most effective exhibit significantly different interaction patterns from the contractor's representatives perceived to be less effective.*

A number of prominent trends emerged from the analyses, which suggest that the communication behaviours of contractors rated most effective are significantly different from those who receive lower ratings.

### *Contractor's representatives' effectiveness: Interaction traits*

Differences were found between the interaction of groups associated with successful outcomes and the communication behaviour of contractor's representatives rated most effective. The findings are modelled in Figure 10.2.

Giving of suggestions, direction and generally showing autonomy, and showing tension were found to be higher in the contractor's representatives' rated the most effective. However, these interaction categories were lower in the groups that resulted in successful outcomes. Thus, a difference exists between the nature of effective contractor's representatives' interaction and the interaction of the successful M&D teams (group interaction).

Contractor's representatives with higher effectiveness ratings use communication acts that demonstrate a high level of self-confidence. Their belief in their own judgement is shown through the amount of direction and suggestions offered to the group. These are the representatives who have historically delivered the most successful projects. The results also indicate that the most effective individuals are more likely to express tension if they are unhappy with a situation and are able to name those they consider to be responsible. In the groups that are associated with successful outcomes, it is normal for these two categories to be lower in successful M&D teams. When the contractor's most effective representatives work in groups associated with successful outcomes, individually they will tend to use greater amounts of suggestions and show tension, although the other group members would limit their use of these categories.

Those considered more effective make use of a broader, more distributed range of communication acts. They are more likely to disagree with others,

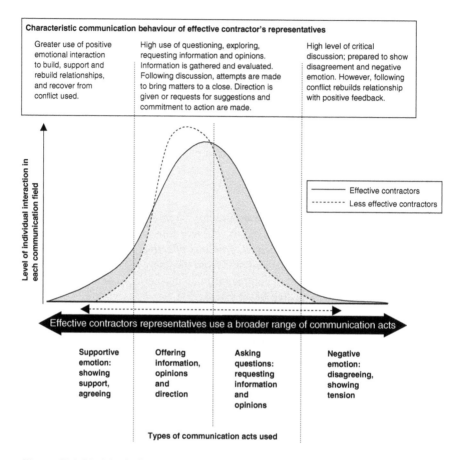

*Figure 10.2* Model of effective contractor's representatives' interaction

put forward suggestions, ask for information, opinions and offer direction than those who are not considered to be so effective. The representatives rated most effective use their communication behaviour within meetings to influence others within the team to adopt practices that result in outcomes most desired by the contracting organisation. By disagreeing more than others, asking for different types of contributions and offering greater direction, they are able to have a greater influence on the group's interaction and behaviour. Thus, their potential to influence the group is increased.

## Limitations and recommendations

The research reported in this book was limited to an investigation of M&D team meetings with a focus on the role of the contractor. The scope of the research was deliberately limited to aspects of interaction that emerged during site-based progress meetings and the differences found between specific project success criteria. This means that the findings only have proven validity within these constraints.

Due to the small sample size in relation to the number of meetings conducted throughout the country, any claims made of its generalisation to the wider population are limited. On the other hand it cannot be said that the data are unrepresentative; a considerable amount of data was collected and analysed. The results of this analysis are indicative of trends within the data and their association with the project that resulted in different outcomes.

The quantitative method used for capturing interaction data also limits the findings of the research. The observations are strictly fixed to the specific definitions of each IPA category as defined by the Bales' protocol. One criticism of the IPA method is that it splits categories into either task- or emotional-based categories: this distinction was sometimes found to be an artificial split as many acts can provide task and emotional information at the same time. Reflecting on the use of the Bales IPA method it is not unreasonable to suggest that the method could be improved (to provide more detailed information) if each act was categorised both on its task dimensions and on emotional intensity. However, where researchers have produced a bespoke method from Bales' IPA their results cannot be compared to other findings based on the original IPA method. It is important that the IPA method is used in its original form; however, other methods can be used to further investigate the individual categories and content of communication. The IPA method works well when used alongside qualitative methodologies. Qualitative data can add considerable detail and description to the otherwise isolated numeric data.

Earlier research on interaction is limited in its relevance to field settings and very little IPA research is available on group interaction in commercial environments. As such it is difficult to determine with any certainty whether or not any of the features identified are peculiar to construction. Observations of other business meetings may find similar tendencies to those reported here. Clearly more research is required in field settings.

### *Recommendations for further research*

There is considerable scope for research in the area of professional group interaction. However, the first recommendation is that this study should be repeated and a confirmatory analysis undertaken to determine whether the findings are applicable to a wider population, for example, in different sized projects. Larger more complex projects with multi-national clients often require a greater number of specialists with different skills.

This initial investigation has identified a number of Bales categories that are associated with project outcomes. Further investigation should be carried out on each of the individual categories to investigate in more detail the nature of the communication acts and how they are used to influence group members' behaviour and actions. Positive and negative emotional interaction is clearly important in organisational communication and needs further investigation. Asking questions and requesting information were found to be important, yet often avoided. Further research is necessary to identify the different ways used to seek information, ideas, opinions and suggestions from others. Also, the relationship between question asking and the fear of being perceived as incompetent, due to an individual's lack of knowledge, understanding or confidence, would benefit from a more detailed research.

Our research did not specifically analyse whether certain patterns of interaction are more common in one of the particular contractual arrangements, but focused on interaction associated with project success from the contractor's perspective. A number of different procurement systems are being developed and some, such as the PPC2000 and the New Engineering Contract (Engineering and Construction Contract), have attempted to improve co-operation. Although earlier research found that contracts have a limited effect on interaction, it is important to determine whether the new systems developed to improve the professional relationships result in different interaction patterns from the traditional contracts. This is an area recommended for future research.

The data on project success were gained from the contractor's organisation. To provide a more holistic view of group dynamics and individual member interaction, the study could be repeated taking other objective stances, collecting the data on project success from the client's, architect's, quantity surveyor's, mechanical and electrical consultant's, professional project manager's and structural engineer's organisations.

The lack of rigorous communication research undertaken during the construction phase means that there remains many aspects that require further investigation. The research that forms the basis for this book looked at the behaviour of participants during the construction phase. Equally important phases dealing with initial team meetings and final meetings were outside the scope of this work. Clearly there is a need for further detailed investigation into construction teams. Work that helps to develop a better understanding of interaction development within temporal construction teams would be particularly welcome. By gathering data of how groups interact, behave and perform (rather than what we would like to think happens), our understanding of groups can be improved. From research appropriate education and training programmes can be developed and improvements may be made.

The findings of this research offer considerable potential benefits to individuals and organisations involved in the construction sector. Practitioners need to be sensitive to a wide range of communication acts. Communication is a critical success factor for all construction projects, regardless of size, scope or complexity. All participants need to be aware of their role in the communication network and their ability to use a wide range of communication acts. Similarly groups should use a full repertoire of communication acts to enable a greater potential to evaluate proposals. In essence we need to (re)value communication in construction teams. From this point onwards we provide some more general reflection and guidance based on our research findings.

## Revaluing communication in construction teams

At the outset of our research we were concerned that interpersonal communication was not sufficiently valued by the actors involved in construction activities. The word communication was used rather loosely in the majority of the literature that we reviewed, usually being used in connection with ICT systems and used to describe the transfer of information, not the exchange of information. Interaction between participants, the manner in which team members communicated with one another to reach mutual understanding, had been largely ignored. Similarly, in practice there appeared to be too little attention paid to the importance of interaction between construction team members and too little importance given to meetings as a forum for integration, knowledge transfer and decision-making. During the course of our research into construction communication we have seen a growing awareness of, and interest in, the softer aspects of communication and in particular the effectiveness of interpersonal communication. One reason for this appears to be associated with the realisation that information management will not, on its own, be an answer to the challenges faced by construction actors. Indeed, the growing interest in knowledge management has shifted focus on individuals and their ability to engage in knowledge sharing activities – this inevitably leads to a focus on individuals' competences and the manner in which they interact. Associated with these trends is an increased interest in, and application of, procurement systems based on relational contracting, for example partnering, strategic alliances etc. These approaches place considerable emphasis on collaborative working and trust, which is built, challenged and redefined through personal interaction and open communication. There is also a growing recognition that management systems, tools and methods are only as good as the people using and implementing them – thus the focus is starting to move towards transferable skills, the competences of the actors and their ability to interact in an effective manner.

The research presented here has focused on the interaction of construction participants in site-based progress meetings. This has helped to highlight the importance of meetings, individual competences and (indirectly) team assembly for construction project success. In an attempt to provide some practical guidance to readers three areas are discussed in more detail:

1. Individual competences (transferable skills and the ability to interact)
2. Team assembly and composition during the life cycle of the project
3. Managing meetings (the forum for interaction).

### *Individual competences and employee suitability*

The relationship between effective individuals and their characteristic inter-action patterns has potential for use in employee selection and development. Group decision-making exercises analysed using IPA techniques could provide a useful assessment tool. Being able to measure and recognise applicants whose interaction behaviour is typical of those professionals perceived as most effective clearly has potential applications. Although effective interaction is only one of many skills required by construction actors, it is clearly an important skill with potential to influence and contribute to group processes.

In addition to selecting new employees, knowing the strengths and weaknesses of those currently employed is essential to organisational and project success. Individuals that possess limited interaction skills can be encouraged to attend training/educational programmes to help develop their ability to interact with other actors. Similarly, those with good interaction skills can serve as role models for others, perhaps even in mentoring roles where appropriate. Video data and self-reflection are useful tools for understanding and developing a broad range of communication approaches, behaviours and techniques. All professionals should be educated and trained to make use of a broader range of interaction behaviour to ensure their contributions have a greater impact on the group. Equally, recognising and being aware of types of interaction that are likely to have a notable influence on the group is a valuable tool. It is clear from the results that emotional interaction plays a significant part in business negotiations. Many professionals have difficulty with emotional exchanges or choose not to use such communication. A lack of emotional awareness, or an inability to use emotional communication, could mean that negotiations are not conducted as successfully as they could be. It is also important that individuals can focus and control their emotions.

The lower rated professionals observed in this study rarely disagreed with other group members. Failing to disagree could mean that proposals are not properly evaluated, thus a weak, probably unworkable, suggestion may be accepted by the team even though a member of the group (who is reluctant to disagree) suspects that the proposal is flawed. Practitioners need to be trained to disagree, engage in conflict and then recover and rebuild relationships following negative emotional encounters. Failing to disagree could mean that valuable time and resources are allocated to unworkable proposals, leading to waste and ineffective project delivery. Similarly, failing to rebuild relationships after particularly tense encounters may be disastrous and could result in conflict escalation, leading to third party intervention and/or legal action.

### Assembling the team

The ability of team members to communicate effectively should, in our opinion, be considered before the choice of procurement route. Emphasis should first be on contact, second on legal contract. The lean philosophy of getting everything right before work starts would tend to support such a view. Adopting a lean approach means that a lot of effort and time is put into the early planning stages to get the structure and processes correct before work starts, once this is complete work commences. In a construction context the effort needs to be given to team assembly. Work into alternative approaches to managing construction projects in Denmark has demonstrated the importance of meetings to discuss and share values before work commences. The creative workshop method being used by a consulting engineer in partnership with large contracting organisations relies on the discussion, sharing and consequent agreement of values (Emmitt *et al.* 2004). A process facilitator, a person with no contractual authority, essentially an informal leader, is brought into the process to work on communication issues and facilitates the creative workshops. Meeting participants are encouraged to discuss and share values; these are then developed into a series of value parameters that form the framework for the stages that follow. All actors sign up to commonly agreed values in line with the partnering philosophy. Although a lot of time and resources are involved in these early meetings research has shown that there are major savings in time and cost further down the line. These savings are related to better communication between actors and much less rework and changes than comparable projects.

Our experience is that team assembly is too often left to chance and/or dealt with too late in the life of the project. Managers should be asking the following questions at the very start of projects:

- Who are the key individuals identified to work on the project?
- Do they have good inter-personal skills? (How was/is this demonstrated?)
- Are the actors socially compatible? (i.e. are they able to work together?)

### *Managing meetings*

The meetings that form the focus of this book were construction progress meetings, which have a specific role in the life of construction projects. Other types of meeting are convened at various stages in projects to achieve quite specific and different objectives. Good project managers are able to adjust their communication skills to suit the purpose of the meeting. Different types of meetings can be used to achieve various objectives, for example team assembly/team building meetings, progress meetings, and impromptu meetings called to solve an unforeseen problem. As seen in our research results, the ability of the project manager to manage the meeting professionally can have a significant effect on the success of the project. These factors are explored in further detail in *Construction Communication* (Emmitt and Gorse 2003).

## Communication as a critical success factor

With the opening up of boundaries in Europe and increased migration of workers internationally (transient labour) there is additional focus on communication (language skills) and skills/competences in developing relations. On a visit to a large construction site in the UK we found work gangs from Germany, Italy and Poland working alongside those from the UK, reflecting an increased trend for worker migration within the European Union. The gangs had been brought into the project because of specific skills shortages within the UK construction industry. The site manager expressed his satisfaction with the quality of the work and the dedication shown by all of nationalities working on the project, while expressing some concern about the workers' inability to understand English. He had devoted extra resources (primarily his own time) to making sure that the work gangs understood their tasks, mainly by spending extra time with the gang leaders. Interpreters (often identified by different coloured helmets or stickers on their safety helmets) are now becoming a common feature on many large construction projects. Further investigation is necessary to identify good communication practices in multinational construction projects, especially given the differences between national cultures and its effect on working and communication practices.

If we are to attain, maintain and improve our communication skills (regardless of the languages we understand) we must work at it. This means investing in education and training programmes for employees, regardless of their position within the organisation. It also means investing in people who are capable of facilitating communication in a project context: people who have the skill to bring teams together and to keep them together during the course of the project. There are examples of a small number of consultants and contractors starting to work with the management of construction projects from a communication perspective and early pilot projects are reporting improvements in project success. Similarly, there are examples of better guidance for project managers that take a softer approach to project management skills. A good example being the Association of Danish Project Managers who have published a guide to help project managers assess their competences (Fangel 2005), and which addresses some of the issues raised in our research.

Communication in construction teams (construction communication) is a way of thinking about the way in which we approach the challenges before us. In many cases it requires a recognition of the importance of communication in project-based organisations and a change in attitude; from defensive and closed communication behaviour to transparent and open exchanges. Parallel to advances in technologies and new managerial approaches to the way in which we organise our construction projects must come improvements in the manner in which we interact with one another. Building on a body of research- and practice-based knowledge about communication in construction teams is one way of improving and enhancing project performance. Systematic research will go some way to helping to identify the challenges ahead, so to will reflection on current working practices.

## Recommended reading

Emmitt, S. and Gorse, C. (2003) *Construction Communication*, Oxford: Blackwell Publishing.

Fangel, M. (ed.) (2005) *Competencies in Project Management: National Competence Baseline for Scandinavia*, Association of Danish Project Management, Hillerød, Denmark.

Fryer, B., Egbu, C., Ellis, R. and Gorse, C. (2004) *The Practice of Construction Management: People and Business Performance*, Oxford: Blackwell Publishing.

# Appendix 1

## Definition of communication acts attributed to Bales' IPA categories

This appendix provides examples of statements that would be associated with each category and it also provides some helpful hints, tips and guidance on the categorisation of communication acts. Some example statements come from Bales (1950), further statements have been added and guidance provided. To ensure consistency observers are strongly advised to consult the original Bales document as well as this document. To create a good understanding of the categories it is suggested that researchers create their own examples of statements that would be associated with each category.

The following information is based on and adapted from Bales (1950).

*1. Shows solidarity – raises others status, gives help, reward*
Acts of affection would include statements or gestures that open or attempt social interaction with a positive statement; for example, to offer a greeting in a friendly manner, saying 'hello', 'hi', 'it's really nice to see you again' or other positive gesture. Showing consideration for the other person, use of first name in a friendly warm manner, for example 'It's Jeff, isn't it?' 'Really good to see you again.' 'It's been a long time, you're looking well'. Offering positive encouraging comments, complimenting behaviour or action, 'that's a good point' 'that's a really good point you've raised', praising, congratulating or showing approval, for example 'that's fine', 'good job', 'well done', 'congratulations', 'that's excellent', 'I couldn't have done that better myself', 'that's much better than I could do', 'how do you do things so quickly', 'you always do a good job', 'I like what you've done' 'The quality of the work is pretty good and you've made good time' 'I'm really pleased with this'. Category 1 can be used in response to category 11 (showing tension) by offering to help; giving resources, assistance, time or money; offering to share, giving support or reassurance; comforting a person. Category 1 can also be used in response to category 12 (showing antagonism) following a period of difficulty, intense discussion, heightened tension or aggression – category 1 act used in response to category 12 include acts of pacification, attempt to allay the opposition, being discreet, tactful, diplomatic, urging unity, suggesting compromise. Support act or befriending, for example

'I can see how you feel', 'I understand the problem', 'can I help?' 'If we pull together we can resolve this', 'let's see if we can help', 'we must be able to resolve this'.

## 2. Shows tension release – jokes, laughs, shows satisfaction

Expression of positive feelings better after a period of tension or negative emotion. Expression of contentment, satisfaction or gratification. Showing sign of heightened positive emotion, enjoyment or enthusiasm. Making friendly jokes (would not include laughing at someone or making fun of a situation). Jokes that attempt to make people smile or acts that help change others to express a more positive demeanour, these could be spontaneous or as a result of an attempt to smooth over some tension. The light hearted statements may reduce the negative intensity, or break up a period of difficult or embarrassing silence. Other encouragements to express positive emotion or release tension would include bantering, kidding, horseplay; the element of friendliness and positive encouragement must be at the fore. Laughing must be sensitive to others otherwise it could antagonise and would be category 12. Category 2 is used to disperse negative emotion, feeling and tension; some acts of humour can be hurtful and would not fall into this category. Statements that patronise or are sarcastic would not fall into this category. There are many different ways that a person can help pacify a situation or relieve tension or encourage the development of a more positive atmosphere. During a period of negative emotion group members may attempt to reduce the tension, sometimes without success; however, such acts should be noted, whether considered successful or not. Although tension may not disperse once such acts are used they may have effected the development or prorogation of greater tension. The focus is on light hearted, humorous or any statement that attempts to lift positive emotion or reduce tension.

## 3. Agrees – shows passive acceptance, understands, concurs, complies

In response to category 1 (showing solidarity) or category 2 (showing tension release) indication of modesty, being respectful and unassertive, for example 'that's what I'll do', 'then I guess we're all agreed on that', 'that's the way I see it too', 'I think you're right about that', 'I agree', 'we could do it that way', 'I suppose that is worth trying' 'Yes, I see what you mean'.

In response to category 3 (agreeing) includes confirmation by repetition or affirmation, or appears to come to a decision. In response to category 4 (gives suggestion), concurrence in a proposed course of action or agreeing to action. In response to asking for or giving suggestion, for example if the actor complies with request, co-operates with order, obliges the other, for example 'I second the motion', 'let's do that then' 'go on then, we'll try it out', 'we can do that'.

In response to giving opinion, agreement with observation, report, analysis or diagnosis for example 'that's the way I see it too' 'I would do that too', 'I think you're right about that', 'yes that's true', 'I can see your point', 'I have the same view'.

In response to category 6 (giving information), recognition of interest, giving signs of attention 'I see', 'oh, I understand', etc. showing comprehension and understanding. Would include agreeing with initial statement, but does not necessarily mean the recipient will agree with any final proposal or suggestion. Participant may accept that they were wrong, for example in response to category 10 (disagreeing) by admitting to an error or oversight, admitting that an objection made by the other was valid. A communication act may fall into this category if it pre-empts a level of dissent and phrases the statement so that it is sensitive to any disagreement. When a person introduces a phrase that anticipates disagreement they may, for example, state, 'I may be wrong about this, but . . .', 'You have good reasons not to agree with this, but I think we need to consider all arguments'. Such statements would commonly have at least two parts to the statement: the first concentrates on passive agreement, being sensitive to the group, and the other parts of the statement may bring new information or a suggestion into the arena. The communication act category 3 would be used to code the passive agreement; the second part of the statement would be coded separately. Category 3 could be used in response to communication act category 11 (shows tension) by showing an indication of permissive attitude, or consenting to a request, or agreeing with the argument or situation. In response to category 12 (shows antagonism), the communication act indicates that the actor is submissive, allows criticism without retaliation, does not argue or show dissent, but passively accepts the situation.

### 4. Gives suggestion – direction, implying, autonomy for others
An act that suggests a concrete way of attaining a desired goal. The act may attack, modify or build on another situation, could be used to provide hypothetical example, situation, exploration or demonstration. The statement could be used to show where, when, how and why. Communication of this nature attempts to guide others. Can be used to prevail upon or persuade others or exhort, urge or inspire. Could be used to recognise and confirm request; however, the person making the suggestion, the person whom conforms and sets the direction must be acting as a legitimate agent for group showing some autonomy. Emotionally toned requests for help are classified in category 11. The emphasis of this communication act is that it is used to establish direction, the person administering the act, assumes autonomy for others, establishes a level of leadership or guardianship over others and attempts to influence the direction of the group. The act is used to provide a decision. The person making the suggestion assume some responsibility, power, authority or attempt to gain the groups recognition so that the suggestion made is given due consideration.

Examples of suggestions:
'We will have to stop at the end of the hour', 'Consider for a moment what would happen if'. 'Suppose we set up the following situation. John, you take the role of the chairman?' 'Go ahead', 'We need to consult a specialist on this matter specialist' 'Right, we will appoint the contractor', 'I suggest we make him redundant', 'We need to move things on, the decision has been made', 'If you have not completed by Tuesday, I will take action', 'You have a specification to work to, all of the work

will be made good', 'Right, I suggest we bring these issues to a close', 'That's the most useful suggestion, we will do that', 'We have to make a decision, so I suggest . . .'

### 5. Gives opinion – evaluation, analysis, express feeling wish

This can be the process and development of action, in its inferential aspects. These can be considered as indicators of the thought and cognition processes that lead to understanding. Rephrasing statements so that others are able to see and explore the other issues associated with the problem. Examples would include providing an insight or giving reasoning; working out problems reckoning or calculating; openly thinking through issues, musing or concentrating; providing an analogy; suggesting cause and effect; breaking issues down, or being intuitive. Category 5 includes expression of value; suggesting an ambition or aspiration; showing determination or courage. Communication acts includes stating or suggesting a desire, want, preference, liking, wishing, hoping, moral obligation, or affirmation of values.

An analysis of the situation could include a statement of policy, outlining guiding principles, or providing a legal perspective which could impinge on a matter, or referring to a broad and indefinite future. The category also includes examination or reflection of own role, attempts to understand and diagnose in an objective way, pointing out or attempting to identify patterns and relationships. Opinions can be used to arrive at new insights.

Following category 6 participants may respond with a category 5 statement such as: 'I wish we could fix it so . . .', 'I think we ought to be fair about this,' 'I believe there are a number of issues to consider', 'I hope we can do something about that', 'That seems to be the right thing to do.' 'I think I can summaries the main issues . . .', 'This is quite difficult we are considering a number of issues which seem to be . . .', 'I think we should consider . . .'.

'I must have been so mad at him that I didn't see he was trying to help me', 'Probably, I don't realise how nervous I am in a situation like that', 'I can see now that I totally misjudged the situation', 'Maybe we got off the track because some of us were more anxious to show what we knew in relation to the problem'.

'According to my calculations it must be about three metres', 'Well let's see, two times the square root of X is', 'It's the same as . . .', 'If you take this dimension and subtract that then . . .' 'If I try to do this a different way . . .' 'If I remove the extra stairwell and store cupboard the floor area the toilets could be accommodated'.

### 6. Gives orientation – information, repeats, clarifies, confirms

This includes a recall of statements, acts which are intended to secure or focus attention on communication; stating someone's name to ensure orientation, clearly identifying the recipient; mentioning the problem to be discussed; restating facts to offer guidance; calling attention to what is going to be addressed and discussed; pointing out the relevance at what an actor is saying; making reference back to agenda; giving an indication that one is entering into a new phase of the discussion; actors simply reporting issues or looking forward. Someone without inference tells about a thought, feeling, action or experience. Only statements that are non-inferential towards self or other, for example showing understanding of another by repeating without inference. Statements must be made without offering extra information or any form or analysis. Includes statements about facts about the nature of the group, issue, objective; such facts must be straightforward, non-inferential, non-emotionally toned and descriptive observations.

Statements that would fall into category 6 would include: 'Ah' (confirming), using a persons name, for example 'Chris, Tom, Alex... this is the drawing I was telling you about', thus acknowledging the person or persons to whom the information will be addressed. 'There are two points I would like to make.' 'In the first place,' 'Now with regard, to our problem of...' 'Going back for a moment' 'What I am about to say relates to...' 'That seems to finish our agenda.' 'We were just discussing...' 'I'll bring you up to date on what we were doing' 'This is the cladding document', 'This is the project schedule', 'The minutes from the last meeting and action points'. 'Let's see what's next on the agenda' 'Item 5.3', 'Referring back to the previous page, the minutes state...'

In response to category 7, category 6 can be used to clarify an issue, for example 'I felt pretty downhearted about that time', 'They all thought I was mad', 'This secretly pleased me. I was actually on the other side', 'I'm thirty three years old', 'I have lived here all my life', 'I'm the architect', 'I'm a qualified mechanical engineer as well as a structural engineer', 'I'll never forget the time I...', 'We were down at the bottom of the site and I pointed out that some of the ridge tiles on the roof were missing' 'These are the colour samples'.

*7. Asks for orientation – information, repetition, confirmation*
Category 7 includes cognitive acts that indicate or express a lack of knowledge sufficient to support action. Participants may show a level of confusion or uncertainty about a position or issue. This act can be used to request clarification or ask another actor to provide further information. Includes an indication that the actor is puzzled. A direct or outright question that requires the giving of factual rather than inferential information would fall into category 7. The information requested needs to be something that can be true or false, would not include opinions or analysis. Includes less focused more indefinite expression of a lack of knowledge or cognitive clarity.

Examples include:
'What?', 'What was that? I didn't quite understand you', 'Could you say that again please' 'Would you repeat that?', 'I don't quite get what you mean', 'Where are we?', 'Are we on item 4.2?', 'Is it the cladding we are talking about?', 'What about the roof detail', 'Where do we stand now?', 'I don't know about this. (I have looked) but I can't make it out', 'It isn't clear to me', 'It may be true or it may not', 'What day of the month is it?', 'What was the date?', 'I'm not sure of the exact date', 'What was the problem?', 'Were the results clear?', 'Who is in charge of the arrangements?', 'Who is the contractor?', 'Who was the project manager on that contract?', 'How long have you lived here?', What's the project duration?', 'Let's see, how old is this document?'

*8. Asks for opinion – evaluation, analysis, expression of feeling*
These include inferential communication acts, aimed at exploring the feelings, values, intentions and inclinations of others. Acts would include questions that allow the actors to express interest or disinterest (not restricted to agreeing or disagreeing or a predetermined answer). Category 8 can be an open-ended question, includes questions and statements that leaves others to provide their opinion, diagnosis or analysis of the situation. The statement or question encourages responses which seek inferential interpretation, hypothesis, diagnosis, summary or further analysis of some idea from the others situation or opinion. Category 8 includes less focused more inferential statements.

Examples of category 8 statements include:
'Tell me more about it', 'You are free to talk about anything you like', 'What were your thoughts?', 'How do you feel about that?', 'What's your view?', 'What do you think?', 'What's your opinion?', 'What is the sense of the meeting?', 'What do you think our policy should be?', 'What do you think we should aim at?', 'Where do you think we are heading?' 'How long do you suppose it would be?', 'I can't figure out how long it would take', 'I wonder what changes, if any, that would involve', 'I don't know whether it would require changes or not', 'Do you think there are any other possibilities?', 'I wonder if we are proceeding in the most effective way', 'I don't know how I really feel', 'How do you feel about this situation?', 'What's your company's view on this?', 'What's your stance?', 'Do you think we need to consult a specialist?', 'Can you give me your professional opinion?', 'How would an Engineer view this?'.

*9. Asks for suggestion – direction, possible ways of action*
The process of requesting a decision, proposal or action. Category 9 includes all questions or requests explicit or implicit, which ask for suggestions as to how the group shall proceed or how issues are to be moved forward. Requests are for concrete ways, means and goals. The emphasis is on searching for the way forward – finding ways, means and solutions to problems. Requests may include suggestions of where to start, what to do next and what to decide. Note that appeals for suggestions that have an emotional undertone should be in category 11, for example 'Gosh, what do I do now? 'So, how do you suggest we sort this out!', 'This is rubbish, what will you do when you're out of work!', 'I'll ask you what you are going to do, but I know you won't give me a proper answer.'

Examples of requests for suggestions include:
'Is there a motion on this point?', 'What should I do next?', 'When will you have the floor finished?', 'What date will you hand that section over?', 'I wonder what we can do about this?', 'I don't know what to do? What do you suggest?', 'What shall we talk about today?', 'How will you resolve this matter?', 'What action are you going to take?', 'Can you suggest a solution?', 'I would like you to propose a solution to this', 'How do we take this forward?', 'When will you have this section finished?'.

Note that some requests may sound like requests for information; however, if the question implies that the respondent(s) may have to undertake some action as a result of the response then it is a request for direction.

### 10. Disagrees – shows passive rejection, formality, withholds help

Includes any indication that a person is rejecting what was being said, for example if a participant expresses an attitude that the observer considers over cool or frigid. Where a person purposely blanks another, this could constitute disagreement. Any situation in which an emotional response would be expected, where the actor refuses to give applause, or is unappreciative, does not acknowledede, is edging, ungrateful, hard to please, or where it is difficult to obtain a positive or supportive reaction. Category 10 includes passive forms of rejection, silent, uncommunicative, inexpressive or impassive responses. Silence should be a failure to respond rather than were a group become quite because member's contributions die out.

Where a person appears reticent, detached, isolated and indifferent, this would fall into this category. If people set themselves up as unapproachable and exclusive this would constitute a category 10 type act. Where a person is more aggressive in the form of disagreement these acts are categorised as 12. In response to category 5 (giving opinion) and category 6 (giving information) category 10 includes milder forms of disagreement, disbelief, astonishment and amazement. Strong statements that carry an emotional tone belong to category 12.

Examples of category 10:
'No, I don't think so', 'I disagree with that', 'If you propose that I will have to stand in opposition', 'If you make those changes I will have to reject the work', 'It seems to me there is more to that. In fact I remember seeing at least five', 'No, I don't think that will work', 'That can't be right', 'I'm not sure about that', The act can be used to offer a refuting statement, for example 'I would do it differently', 'I wouldn't do it like that', 'You have failed to take all of the main issues into account', 'You can't do that without considering the other issues', 'We won't agree on this', 'It's not going to work like that'.

### 11. Shows tension – asks for help, withdraws out of field

Where those interacting show that they are on edge, keyed up, agitated, startled, disconcerted, alarmed, dismayed, concerned, or express fear, apprehension, worry, are overcautious, over wary, evade an issue, are tense, over prudent, or are conscientious because of fear of provoking opposition of hostility, this would constitute a category 11 act. Where a person is blocking something out, or bottling things up or has misgiving about something that she/he has done, then this is also a category 11 type act. If the actor admits shame or guilt or humiliates themselves then this could be signs of tension. Category 11 also includes, being apologetic, remorseful, self-dissatisfied, critical, unrepentant, depreciating, accusing, exposing, conviction, condemning and humiliating. Showing expression of frustration of feeling towards one's own efforts (dissatisfied, disappointed and displeased) and these feelings are expressed only in a diffuse way, with no special social object of indication then they are scored under category 11. Requisitions for permission or help that carry a noticeable undertone of emotionality are included. Where a person withdraws out of the field would be included. Category 11 also includes any behaviour that indicates to the observer that the actor is inattentive, bored or psychologically withdrawn from the problem. Includes yawning, unaware, leaving, quitting, retreating or negatively toned evaluations of self-conduct. Where a person indicates that they are irritated with the discussion, or shows that they would rather be somewhere else then this could constitute a category 11 act.

### 12. Shows antagonism – deflates others status, defends or asserts self

These are self-defensive statements. Category 12 would include autocratic control, attempts to control, regulate, govern, direct or supervise in a manner which the observer interprets as arbitrary or autocratic, in which freedom of choice or consent for the other person is either greatly limited or non-existent. Restricting the other's power by demands or commands such as 'hurry up', 'stop that', 'come here' are category 12 acts if stated with an emotional undertone. An emotive or emotional act that points, pushes, pulls or otherwise directs an actor. Where an actor forcefully attempts to lay down principles of conduct, standards, laws, attempts to judge, give decision, compel, master, dominate, gives warning, gives threats. Category 12 act attempt to exert influence and control over others without any emotional restraint. Such acts may be used to forcefully exert an influence on others, reject others or escape. Includes acts that are negative stubborn and resistant. Includes act to shake off restraint or get free. Any behaviour that circumvents authority, shows independence, non-conformity, disobedient, non-compliant, insubordinate, rebellious, wilful, nagging, badgering, harassing or annoying. An actor who is flippant and unrepentant when justly accused would be using a category 12 type act. Communication acts that are status deflating, used to interrupt and interfere with another actor or helping when someone clearly does not want help, can be antagonistic. Acts that are used to belittle, are depreciating, minimising, make fun of an actor are category 12 acts. Includes acts of status seeking, being self assertive, acts that are used to impress the other with his or her importance. Where the act dramatises herself or himself, an act used to show off, give expressive gestures with a negative undertone, if speech is extravagant or used to brag, then this constitutes a category 12 act. Other negative emotional reactions include aggressive, combative, quarrelsome, argumentative or challenging behaviour.

Examples of type 12 acts:
'I have been in this industry for 30 years, you know nothing son!', 'At what point did I say you could do that!', 'You clearly have no idea!', 'When you've been doing this as long as me you'll know what I'm talking about', 'You have absolutely no idea', 'When will you learn?', 'Next time think before opening your mouth', 'The world would be a better place without you', 'It's clear that you do not understand the importance of this project', 'Don't be a fool', 'Anymore comments like that and I will leave this meeting', 'Just try it!', 'The standard of your work is

appalling'. 'O.K.! So exactly when do you intend to start work on this?' 'If you give this project such a low priority, then maybe we should appoint another contractor', 'It's about time we saw some action from your team!' 'What is your problem?',

'I'm not working on this anymore!', 'I've had enough of this project, you can do it yourself', 'Right I'm leaving, cope on your own', 'Up yours!', 'Your really are an idiot'.

## Guidance on attributing communication acts to categories

As observers become more acquainted with the Bales IPA system, they should rely on their own intuitive judgement. As recipients to the communication that takes place, observers are sensitive to emotion. When an observer suspects that an act carried a emotional undertone, was used sarcastically or was enhanced with positive emotion then, more often than not it was delivered with emotion. Our ability to communicate is dependent on our ability to distinguish different expressive and emotional signals, some signals are very subtle, yet we can still recognise them. The tone of voice, pitch, timbre, posture, body language and facial expression all provide clues. Normally during conversation many of these signals are processed subconsciously and the mind quickly and subconsciously processes a response. However, when observing, we are not interested in taking part in the conversation and are solely dedicated to identify communication acts and recognising the clues that tell us what these signals are. Through training, observers become more sensitive to subtle signals that help categorise interaction. Observers should allow themselves to pick up and record emotions even the slightest of emotion, which potential changes the meaning or type of communication act.

Two rules which Bales uses to help ensure consistency of coding are:

> *View each act as a response to the last act of the last other, or as an anticipation of the next other.* (Bales 1950: 91)

[and to improve sensitivity]

> *Favour the category more distant from the middle. Classify the act in the category nearer the top or bottom of the list.* (Bales 1950: 93)

Another rule that makes the observer more sensitive to the directive quality of the interaction, is to identify the 'active outgoing emotion' (Bales 1950). The observer should use his/her own intuitive emotions to classify the interaction, based on prior, present and possible future interaction, rather than pure logic that may ignore previous interactions (Bales 1950).

Critics of the Bales system suggest that it is not possible to accurately code interaction; however, the system merely constitutes a second language. Many people are bilingual, and can understand and converse in a number of different languages, few would suggest that this in not possible. In some languages greater attention is given to emotion or intonation and when learning the language consideration is given to such subtle differences. With practice and time we can learn a second language. The Bales IPA system relies on a few basic rules and only uses 12 codes. With minimum training the system can be used with accuracy.

Observers should not be worried about a few errors in coding. When coding short conversations hundreds of communication acts are used, and a small number of errors make little difference to the aggregated data produced by an individual or group. Video data can be reviewed and coded by multiple observers, any differences in coding can be explored. When video data or audio data is used to examine short interaction sequences then the classification and diagnosis of the interaction data is more intense, focused and detailed. Where large samples of interaction are coded, the few coding mistakes made make little difference. When observations are short and video or audio recorded the accuracy of the coding may be increased.

## Example of coded interaction

Each sentence, or speech act, is categorised in brackets at the end of the sentence.

For example
<person speaking> *'this is how information is coded . . .'* (code of communication act)
<Project manager> If we can return to the agenda (4), item 5.1 roof purlins (6). Ian (6) I sent you a letter regarding the deflection in the zed purlins last week and I still haven't had a reply to this (6).
<Structural engineer> I thought I mentioned that at the last meeting (5); the design is correct (6).
<Project manager> The deflection does seem excessive (5), are you sure it's ok (8)?
<Structural engineer> I've checked through my calculations (6).
<Project manager> Are you sure you have allowed for the air handling ducting (7)?
<Structural engineer> Yes (3), my calculations are here (6).
<Project manager> The deflection must be outside the design limits (7), it doesn't looks safe (10). Can we have another look at it after this meeting and think about putting some extra supports in (4).
<Structural engineer> OK (3).
<Project manager> Item 5.2 (6).
<Mechanical and electrical consultant> That's my section. I've nothing major to report (6).
<Project manager> I'm not sure about that, you are three weeks late on the plant details (11).

# Appendix 2

## Observer reliability: Test results

The main researcher and an assistant researcher followed the IPA protocol developed by Bales (1950) and then simultaneously coded a video of group inter-action. At intermediate stages in the training, process tests were carried out to check the degree of agreement between the two coders. Where differences were found between the observations these were discussed and compared against the Bales' protocol.

### Initial reliability and agreement test with one other observer, researcher A

The following Chi-square statistics show an example of test results that indicated a lack of agreement on IPA categories used.

*Agreement test undertaken when observing a four-person group*

Level of agreement found when identifying person speaking:

$\chi^2 = 2.082, \ df = 3, \ p = 0.556$

Level of agreement found when identifying person receiving:

$\chi^2 = 2.018, \ df = 3, \ p = 0.569$

Level of agreement found when classifying communication acts using Bales' interaction process analysis categories:

$\chi^2 = 14.648, \ df = 7, \ p = 0.04$

Bales suggested that the Chi-square statistic at levels above $p = 0.5$ should be used as a measure of agreement rather than high levels of correlation. This is because an acceptable level of correlation can be found when the results from the Chi-square would indicate relatively large differences. Some studies would have found the following correlation (Table A2.1), from the same reliability test, to be acceptable, whereas the previous Chi-square results would indicate acceptable levels of agreement have not been reached.

*Table A2.1* Correlation results for reliability test

| | | Symmetric measures | | | |
|---|---|---|---|---|---|
| | | Value | Asymp. std. error[a] | Approx. T[b] | Approx. sig. |
| Interval by interval | Pearson's R | −0.075 | 0.116 | −0.628 | 0.532[c] |
| Ordinal by ordinal | Spearman Correlation | −0.035 | 0.120 | −0.294 | 0.769[c] |
| No. of valid cases | | 72 | | | |

Notes
a   Not assuming the null hypothesis.
b   Using the asymptotic standard error assuming the null hypothesis.
c   Based on normal approximation.

## Final reliability and agreement test with researcher A

The following provide test results that indicated an agreement on the observations of the person sending messages, receiving messages and the IPA categories attributed (Figure A2.1).

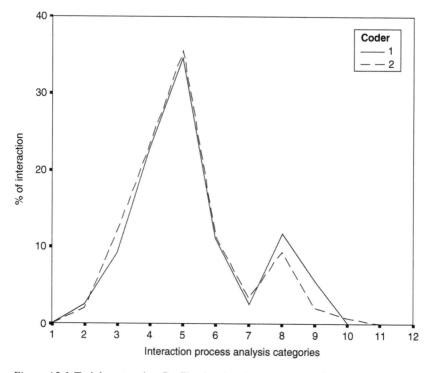

*Figure A2.1* Training exercise: Profile showing the percentage of Bales' IPA categories coded by each observer during simultaneous observation of group interaction

*Agreement test undertaken when observing a three-person group*

Level of agreement found when identifying person speaking:

$\chi^2 = 0.911,\ df = 2,\ \rho = 0.634$

Level of agreement found when identifying person receiving:

$\chi^2 = 1.407,\ df = 3,\ \rho = 0.704$

Level of agreement found when classifying communication acts using Bales' interaction process analysis categories:

$\chi^2 = 4.916,\ df = 8,\ \rho = 0.760$

Chi-square is above the $\rho = 0.5$ level suggested by the Bales method and is taken as reliable.

Following the initial training exercise with one assistant it was decided to repeat the training with two new researchers.

## Initial reliability and agreement test with researcher B and C (agreement between three observers)

The results and graph below show the initial level of agreement between all three researchers when observing a real group, not a video observation (Figure A2.2).

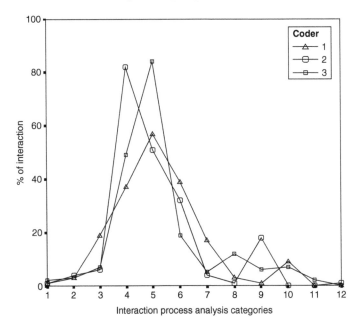

*Figure A2.2* Training exercise: Profile showing the percentage of Bales' IPA categories coded by each observer during simultaneous observation of group interaction; Level of agreement unsatisfactory

*Agreement test undertaken when observing a seven-person group*

Level of agreement found when identifying person speaking:

$x^2 = 12.898$, $df = 12$, $\rho = 0.376$

Level of agreement found when identifying person receiving:

$x^2 = 145.784$, $df = 16$, $\rho = {<}0.001$

Level of agreement found when classifying communication using Bales' interaction process analysis categories:

$x^2 = 103.535$, $df = 22$, $\rho = {<}0.001$

Following the observation of the real group, further training exercises were undertaken. The three observers simultaneously coded a video of group interaction and discussed differences found and compared them against the Bales protocol.

## Intermediate reliability and agreement test with researcher B and C (agreement between three observers)

The initial observations of the video produced results that showed a lack of agreement on the classification of communication acts using Bales' IPA.

*Agreement test undertaken when observing a three-person group*

Level of agreement found when identifying person speaking:

$x^2 = 1.299$, $df = 2$, $\rho = 0.522$

Level of agreement found when identifying person receiving:

$x^2 = 1.299$, $df = 2$, $\rho = 0.522$

Level of agreement found when classifying communication using Bales' interaction process analysis categories:

$x^2 = 27.572$, $df = 12$, $\rho = 0.006$

Although the results between all three observers were not consistent, two of the researchers were in agreement with each other and the protocol. The third researcher found greater difficulty in following the methodology and required further training. The results below present the Chi-square statistic for the two researchers whose level of agreement on the person speaking, person receiving communication and the IPA category was acceptable.

## Reliability and agreement test with researcher with examination between researcher B and main research

*Agreement test undertaken when observing a three-person group*

Level of agreement found when identifying person speaking:

$\chi^2 = 0.227, \ df = 1, \ \rho = 0.634$

Level of agreement found when identifying person receiving:

$\chi^2 = 0.227, \ df = 1, \ \rho = 0.634$

Level of agreement found when classifying communication using Bales' interaction process analysis categories:

$\chi^2 = 1.395, \ df = 6, \ \rho = 0.966$

Even when the level of agreement between all three observers was considered acceptable, differences were noted (Figure A2.3). These were mainly attributable to one of the observer's difficulty with the Bales' protocol. Further training was undertaken.

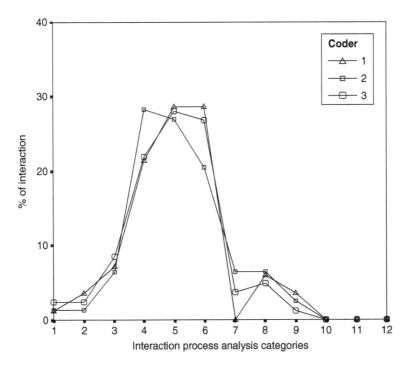

*Figure A2.3* Training exercise: Profile showing the percentage of Bales' IPA categories coded by each observer during simultaneous observation of group interaction; Level of agreement satisfactory

# Appendix 3

## Data collection sheet

The following data sheet was used to record interaction during the group meetings (Figure A3.1). The researcher observing the meeting completed the information at the top of the sheet. The start time of the meeting was recorded, the time when each sheet was completed was recorded at the foot of the page, and the finish time of the meeting was recorded on the last page. While this provided the amount of interaction per time period, it was also used to ensure the sheets were kept in order, the sheets were also numbered.

During interaction a very brief note is made in the 'Brief description' column. This was a maximum of two or three words that indicated the subject being discussed, or cross-referenced the discussion to a section in the minutes or agenda of the meeting. This information was only entered when the topic of discussion changed, usually one or two entries per page were made, depending on how often the topic of discussion changed. Where an agenda was used the agenda item number was used.

Every time a communication act was observed three factors were identified: the person talking, the person who the communication was aimed at and the classification of interaction. The simplest way to record these occurrences was with a dash in the appropriate cell. Occasionally, two or more people from the same professional group listed attended the meeting. To distinguish between the professionals of the same group a '1' was used to identify the person nominated as the first participant in that group and a '2' in the same column was used to indicate the second participant of the group. If 3 or 4 professionals from the same group were present, 3's and 4's would be used. The only rule when using this system was that the numbers that were used to identify the individual professional were maintained throughout all of the meetings of that particular group.

Alternatively when training or studying small groups it may be easier to use the Quantitative Analysis and Direction (QuAD) sheet (Figure A3.2).

| Reference to agenda or problem | Person Speaking - where two or three parties use 1, 2, 3 etc. to identify person speaking | Person being addressed | Bales' IPA |
|---|---|---|---|

Person Speaking columns: Architect | Contractor | Client | Mechanical Engineer | Structural Engineer | Sub contractor | Other | Other (a | c | cl | m | se | sc | o1 | o2)

Person being addressed columns: Architect | Contractor | Client | Mechanical Engineer | Structural Engineer | Sub contractor | Other | Other | Group (a | c | cl | m | se | sc | o1 | o2 | G)

Bales' IPA columns:
1 show solidarity, raises others status, gives help, reward
2 shows tension release, jokes, laughs, shows satisfaction
3 agrees, shows passive acceptance, understands,
4 gives suggestion, direction, implying, autonomy for others
5 gives opinion, evaluation, analysis, express feeling, wish
6 gives orientation, information, repeats, clarifies, confirms
7 asks for orientation, information, repetition, confirmation
8 asks for opinion, evaluation, analysis, expression of
9 asks for suggestion, direction, possible ways of action
10 disagrees, shows passive rejection, formality withholds
11 shows tension, asks for help, withdraws out of field
12 shows antagonism deflates others status, asserts self

Rows numbered 1 through 33, each with:
Person Speaking: a c cl m se sc o1 o2
Person being addressed: a c cl m se sc o1 o2 G
Bales' IPA: 1 2 3 4 5 6 7 8 9 10 11 12

Time when this sheet was completed _____ Sheet number _____

*Figure A3.1* Data collection sheets

| | | Date | Communication – Quantitative Analysis and Direction (QuAD) Tick Sheet | | | | | | | | | | | | | Sheet No. ____ |
|---|---|---|---|---|---|---|---|---|---|---|---|---|---|---|---|---|
| | | No speaker | Sender | | | | | | Receiver | | | | | | | Cross ref |
| | IPA | | a | b | c | d | e | f | a | b | c | d | e | f | |
| 1 | | | | | | | | | | | | | | | | |
| 2 | | | | | | | | | | | | | | | | |
| 3 | | | | | | | | | | | | | | | | |
| 4 | | | | | | | | | | | | | | | | |
| 5 | | | | | | | | | | | | | | | | |
| 6 | | | | | | | | | | | | | | | | |
| 7 | | | | | | | | | | | | | | | | |
| 8 | | | | | | | | | | | | | | | | |
| 9 | | | | | | | | | | | | | | | | |
| 10 | | | | | | | | | | | | | | | | |
| 11 | | | | | | | | | | | | | | | | |
| 12 | | | | | | | | | | | | | | | | |
| 13 | | | | | | | | | | | | | | | | |
| 14 | | | | | | | | | | | | | | | | |
| 15 | | | | | | | | | | | | | | | | |
| 16 | | | | | | | | | | | | | | | | |
| 17 | | | | | | | | | | | | | | | | |
| 18 | | | | | | | | | | | | | | | | |
| 19 | | | | | | | | | | | | | | | | |
| 20 | | | | | | | | | | | | | | | | |
| 21 | | | | | | | | | | | | | | | | |
| 22 | | | | | | | | | | | | | | | | |
| 23 | | | | | | | | | | | | | | | | |
| 24 | | | | | | | | | | | | | | | | |
| 25 | | | | | | | | | | | | | | | | |
| 26 | | | | | | | | | | | | | | | | |
| 27 | | | | | | | | | | | | | | | | |
| 28 | | | | | | | | | | | | | | | | |
| 29 | | | | | | | | | | | | | | | | |
| 30 | | | | | | | | | | | | | | | | |
| | | Start time: | | | | | | | Finish time: | | | | | | | |

*Figure A3.2* QuAD – Quantitative analysis and direction tick sheet

# Appendix 4

## Number of meetings observed

To ensure that the aggregated profile of all the group meetings was not skewed by the numbers of observations, a graph was produced to show the differences in data from all ten projects observed three times and projects observed four times.

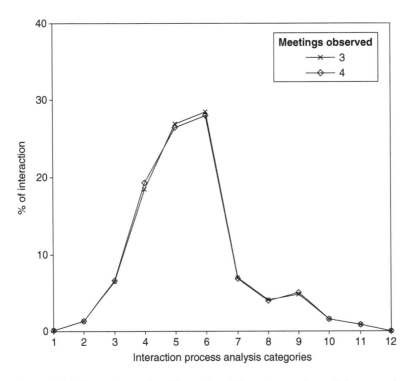

*Figure A4.1* Comparison of profile produced from observations of three meetings with profile produced from all meetings observed (3 and 4 meetings)

# Appendix 5

## Communication models developed from this research

Two models have been developed from our research. The first model attempts to illustrate conflict management (Figure A4.1) and the second addresses the communication attributes of group processes and development (Figure A5.1).

As the group explores issues differences manifest. The differences are a result of different perspectives, beliefs, values and goals. Such differences are to be expected and often result from different education, experiences, expectations, internal or external pressures. As conflict manifests it will either develop into a full dispute that threatens relationships or it will be managed and the group relationship will be sustained. To manage conflict participants must explore the differences looking for a resolution, while at the same time ensuring that the relationship is maintained. Criticism, disagreements and conflict causes tension, this tension should be reduced by communication acts that specifically attempt to reduce the effects of negative statements. Communication acts that are supportive, offer positive emotion or tension release and help maintain relationships throughout the conflict experience. Often those engaging in conflict are preoccupied with the issue and tension and it falls to the other members of the group to offer support and change the atmosphere in the group. After the conflict episode it is important to re-establish relationships. Light-hearted interjections or supportive communication acts help to encourage those previously engaged in conflict to exchange neutral communication. The exchange of communication acts that do not have negative socio-emotional undertones may be the first step in ensuring that work relationships can continue. Where relationships are not managed conflict can escalate, becoming increasingly circular and destructive. Failing to re-establish basic non-emotive communication may result in parties continuing with their conflict and formalising it into a legal dispute. Both the parties engaged in conflict and others can be helpful in pacifying emotions, reducing tension and resuming some normality to the group.

The research has shown that Bales (1950) task and social dimensions are fundamental to the study of work-group behaviour. Work-groups cannot function without building and sustaining relationships that are capable of achieving the group task. The interaction within the group is focused on achieving the task,

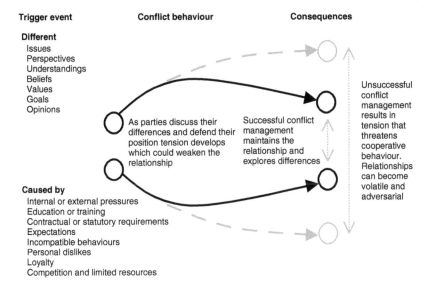

**Trigger event**       **Conflict behaviour**       **Consequences**

**Different**
Issues
Perspectives
Understandings
Beliefs
Values
Goals
Opinions

As parties discuss their differences and defend their position tension develops which could weaken the relationship

Successful conflict management maintains the relationship and explores differences

Unsuccessful conflict management results in tension that threatens cooperative behaviour. Relationships can become volatile and adversarial

**Caused by**
Internal or external pressures
Education or training
Contractual or statutory requirements
Expectations
Incompatible behaviours
Personal dislikes
Loyalty
Competition and limited resources

*Figure A5.1* Model of conflict management

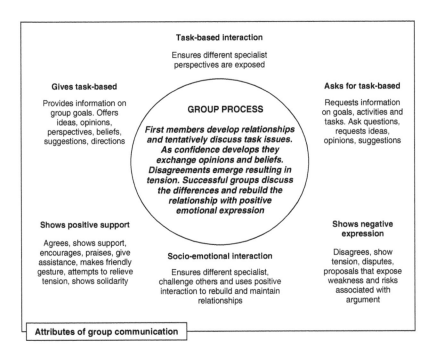

**Task-based interaction**

Ensures different specialist perspectives are exposed

**Gives task-based**

Provides information on group goals. Offers ideas, opinions, perspectives, beliefs, suggestions, directions

**GROUP PROCESS**

*First members develop relationships and tentatively discuss task issues. As confidence develops they exchange opinions and beliefs. Disagreements emerge resulting in tension. Successful groups discuss the differences and rebuild the relationship with positive emotional expression*

**Asks for task-based**

Requests information on goals, activities and tasks. Ask questions, requests ideas, opinions, suggestions

**Shows positive support**

Agrees, shows support, encourages, praises, give assistance, makes friendly gesture, attempts to relieve tension, shows solidarity

**Socio-emotional interaction**

Ensures different specialist, challenge others and uses positive interaction to rebuild and maintain relationships

**Shows negative expression**

Disagrees, show tension, disputes, proposals that expose weakness and risks associated with argument

**Attributes of group communication**

*Figure A5.2* Communication attributes of group process and development

yet the task often relies on multiple contributions from members. To achieve the group goal relationships must formed and sustained. Interpersonal and group relationships are formed and managed through interaction. Communication acts that have task and social attributes are used to form the group's co-operative and regulatory framework that decides who and how much assistance is given to tasks. The task and social dimensions of group interaction are interrelated and often intermingled. Figure A5.2 provides some common attributes of group development.

# Bibliography

Abadi, M. (2005) *Issues and Challenges in Communication Within Design Teams in the Construction Industry*. PhD Thesis, Faculty of Engineering and Physical Sciences, School of Mechanical, Aerospace and Civil Engineering, The University of Manchester.

Ackoff, R. (1966) *Structural Conflict Within Organizations in Operational Research and the Social Sciences*, ed. J.R. Lawrence. London: Tavistock Publications.

Ahmed, S.M. and Kangari, R. (1995) Analysis of client satisfaction factors in the construction industry. *Journal of Management in Engineering*, **11** (2), 36–44.

Albrecht, T.L. (1997) Patterns of socialisation and cognitive structure, in: G. Philipsen and T.L. Albrecht (eds) *Developing Communication Theories*. New York: State University Press, 7–28.

Allinson, C.W. and Hayes, J. (1996) The cognitive style index: A measure of intuition-analysis for organisational research, *Journal of Management Studies*, **33**, 119–135.

Alrugaib, T.A. (1995) *Project Information, Office Automation, and Quality in Building Production Process in Saudi Arabia*. PhD thesis, Cardiff University.

Anderson, C.M. and Martin, M.M. (1995) The effects of communication motives, interaction involvement, and loneliness on satisfaction: A model of small groups, *Small Group Research*, **26**, 118–137.

Anderson, C.M., Riddle, B.L. and Martin, M.M. (1999) Socialization processes in groups, in: L.R. Frey (ed.) *The Handbook of Group Communication Theory and Research*. London: Sage Publications, 139–163.

Aries, E. (1976) Interaction patterns and themes of male, female and mixed groups, *Small Group Behaviour*, **7**(1), February.

Armstrong, S.J. and Priola, V. (2001) Individual differences in cognitive styles and composition of self managing work teams, *Small Group Research*, **32** (4), 283–312.

Arnold, J., Cooper, C.L. and Robertson, I.T. (1995) *Work Psychology, Understanding Human Behaviour in the Workplace*, 2nd edn. London: Pitman Publishing.

Atkinson, P. and Hammersley, M. (2001) Ethnography and participant observation, in: D. Silverman (ed.) *Interpreting Qualitative Data*. London: Sage Publications.

Austin, S., Baldwin, A. and Newton, A. (1993) Modelling design information in a design and build environment, *Association of Researchers in Construction Management*, 9th Annual Conference, 14–16 September, Oxford University, 73–84.

Averill, J.R. (1993) Illusions of anger, in: R.B. Felson and J.T. Tedeschi (eds) *Aggression and Violence: Social Interaction Perspectives*. Washington: American Psychological Association, 171–192.

Azam, M.A.M., Ross, A.D., Fortune, C.J. and Jaggar, D. (1998) An information strategy to support effective construction design decision making, in: W. Hughes (ed.) *14th*

*Annual ARCOM Conference*, 9–11 September 1998, University of Reading, Association of Researchers in Construction Management, Vol. 1, 248–257.

Bakeman, R. and Gottman, J.M. (1997) *Observing Interaction: An Introduction to Sequential Analysis*, 2nd edn. Cambridge: Cambridge University Press.

Bales, R.F. (1950) *Interaction Process Analysis: A Method for the Study of Small Groups*. Cambridge, USA: Addison-Wesley Press.

——(1953) The equilibrium problem in small groups, in: T. Parsons, R.F. Bales and E.A. Snils (eds) *Working Papers in the Theory of Action*. New York: Free Press, 111–163.

——(1958) Task roles and social roles in problem solving groups, in: E.E. Maccoby, T.M. Newcomb and E.L. Hartley (eds) *Readings in Social Psychology*, 3rd edn. New York: Holt, 437–447.

——(1970) *Personality and Interpersonal Behaviour*. New York: Holt, Rinehart and Winston.

——(1980) *SYMLOG Case Study Kit: With Instructions for a Group Self Study*. New York: Free Press.

Bales, R.F. and Hare, A.P. (1965) Diagnostic use of the interaction profile. *Journal of Social Psychology*, **67**, 239–258.

Bales, R.F. and Slater, P.E. (1955) Role differentiation in small decision-making groups, in: T. Parson and R.F. Bales (eds) *Family Socialization and Interaction Process*. Glencoe, Illinois: Free Press, 259–306.

Bales, R.F. and Strodtbeck, F.L. (1951) Phases in group problem solving, *Journal of Abnormal and Social Psychology*, **XLVI** (46), 485–495.

Bales, R.F., Strodtbeck, F.L., Mills, T.M. and Roseborough, M.E. (1951) Channels of communication in small groups, *American Sociological Review*, **16**, 461–468.

Bales, R.F., Cohen, S.P. and Williamson, A. (1979) *SYMLOG: A System for the Multiple Level Observation of Groups*. New York: Free Press.

Ball, M. (1994) Vacillating about Vietnam: Secrecy, duplicity, and confusion in the communication of President Kennedy and his advisors, in: L.R. Frey (ed.) *Group Communication in Context: Studies of Natural Groups*. Hove: Lawrence Erlbaum Associates, 181–189.

Banwell Report (1965) *The Placing and Management of Contracts for Building and Civil Engineering Works*, HMSO.

Barge, J.K. and Keyton, J. (1994) Contextualizing power and social influence in groups, in: L.R. Frey (ed) *Group Communication in Context: Studies of Natural Groups*. Hove: Lawrence Erlbaum Associates, 85–105.

Belbin, R.M. (1981) *Management Teams: Why the Succeed or Fail*. London: Heinemann.

——(1993) *Team Roles at Work*. Oxford: Butterworth-Heinemann.

——(2000) *Beyond the Team*. Oxford: Butterworth-Heinemann.

Bell, J. (1999) *Doing Your Research Project: A Guide for First-time Researchers in Education and Social Science*. Bristol: Open University Press.

——(2001) Patterns of interaction in multidisciplinary child protection teams in New Jersey, *Child Abuse and Neglect*, **25**, 65–80.

Bellamy, T., Williams, A., Sher, W., Sherratt, S. and Gameson, R. (2005) Design communication: Issues confronting both co-located and virtual teams, in: F. Khosrowshahi (ed.) *21st Annual ARCOM Conference*, 7–9 September, SOAS, University of London, Association of Researchers in Construction Management, Vol. 1, 353–361.

Bemm, D.J., Wallach, M.A. and Kogan, N. (1970) Group decision making under risk of aversive consequences, in: P. Smith (ed.) *Group Processes Selected Readings*. Middlesex: Penguin, 352–366.

Bentley, T. (1994) Facilitation: Providing opportunities for learning. *Journal of European Industrial Training*, MCB University Press Limited, **18** (5), 8–22.

Berelson, B. (1971) *Content Analysis*. London: Hafner.

Berger, P.L. (1986) *Invitation to Sociology a Human Perspective*. Middlesex: Penguin Books.

Bernard, H.R., Killworth, P.D. and Sailer, L. (1982) Informant accuracy in social network data V's an experimental attempt to predict actual communication from recall data, *Social Science Research*, **11**, 30–66.

Betney, T. (1997) Poor communication causes defects, *Construction Manager*, June, **3** (5), 5.

Bettenhausen, K. and Murnighan, J.K. (1985) The emergence of norms in competitive decision-making groups, *Administrative Science Quarterly*, **30**, 350–372.

Bettman, J.R. and Park, C.W. (1979) *Description and examples of a protocol coding scheme for elements of choice processes*, Working paper No. 76. U.C.L.A., Centre for Marketing Studies, Los Angeles.

Billingsley, J.M. (1993) An evaluation of the functional perspective in small group communication, in: S.A. Deetz (ed.) *Communication Yearbook*, **16**. Newbury Park, California: Sage, 601–614.

Borgatta, E.F. and Bales, R.F. (1953) Task and accumulation of experience as factors in the interaction of small groups. *Sociometry*, **16**, 239–252.

Bormann, E.G. (1996) Symbolic convergence theory and communication in group decision making, in: R.Y Hirokawa and M.S. Poole (eds) *Communication and Group Decision Making*, 2nd edn, London: Sage, 81–113.

Bowditch, J.L. and Buono, A.F. (1990) *A Primer on Organizational Behaviour*, 2nd edn. New York: John Wiley & Sons.

Bowen, P.A. (1993) *A Communication Based Approach to Price Modelling and Price Forecasting in the Design Phase of the Traditional Building Procurement Process in South Africa*. PhD thesis, Department of Quantity Surveying, University of Port Elizabeth.

——(1995) *A Communication Based Analysis of the Theory of Price Planning and Price Control*. London: RICS research paper, **1** (2).

Bowen, P.A. and Edwards, P.J. (1994) A communication based evaluation of cost planning: Cost data considerations, *Association of Researchers in Construction Management, 10th Annual Conference*, 14–16 September, Loughborough University of Technology, 381–391.

——(1996) Interpersonal communication in cost planning during the building design phase, *Construction Management and Economics*, **14**, 395–404.

Bowley, M. (1960) *Innovations in Building Materials: An Economic Study*. London: Gerald Duckworth & Co.

——(1966) *The British Building Industry: Four Studies in Response and Resistance to Change*. Cambridge, UK: Cambridge University Press.

Boyd, D. and Pearce, D. (2001) Implicit knowledge in construction professional practice, *Association of Researchers in Construction Management, 17th Annual Conference*, 5–7 September, University of Salford, 37–46.

Brown, R. (2000) *Group Process: Dynamics Within and Between Groups*, 2nd edn. Oxford: Blackwell.

Brownell, H., Pincus, D., Blum, A., Rehak, A. and Winner, E. (1997) The effects of right hemisphere brain damage on patients', Use of terms of personal reference, *Brain and Language*, **57**, 60–79.

Bryman, A. (1988) *Doing Research in Organizations*. London: Routledge.

——(1992) *Research Methods and Organisation Studies*. London: Routledge.

Building Industry Communications (1966) *Interdependence and Uncertainty: A Study of the Building Industry*. London: Tavistock.

Building Research Establishment (1987) *In Achieving Quality on Building Sites*. London: NEDO Report.

Burgoon, M., Humsaker, F.G. and Dawson, E.J. (1994) *Human Communication*, 3rd edn. London: Sage.

Burke, P.J. (1974) Participation and leadership in small groups, *American Sociological Review*, **39**, December, 832–843.

Byrd, T. (1998) Taywood targets zero defects, *Construction Manager*, **4** (3), April, 4–5.

Calvert, R.E., Bailey, G. and Coles, D. (1995) *Introduction to Building Management*, 6th edn. Oxford: Laxton's.

Campbell, A.C. (1968) Selectivity in problem solving, *American Journal of Psychology*, **81**, 543–550.

Cappella, J.N. (1997) The development of theory about automated patterns, *Developing Communication Theories*, New York: State University Press.

Capers, B. and Lipton, C. (1993) Hubble space telescope disaster, *Academy of Management Review*, **7** (3), 23–27.

Cartwright, D. and Zander, A. (eds) (1968) *Group Dynamics: Research and Theory*. 3rd edn. New York: Harper & Row.

Cassell, C. and Symon, G. (1994) *Qualitative Methods in Organizational Research*. London: Sage.

Charoenngam, C. and Maqsood, T. (2001) A qualitative approach in problem solving process tracing of construction site engineers, *Association of Researchers in Construction Management, 17th Annual Conference*, 5–7 September, University of Salford, 475–483.

Cheek, F.E. (1965) Family interaction patterns and convalescent adjustment of the schizophrenic, *Archives of General Psychiatry*, **13**, 136–145.

Cherns, A.B. and Bryant, D.T. (1984) Studying the client's role in construction management, *Construction Management and Economics*, **9**, 327–342.

Chou, C. (2002) A comparative content analysis of student interaction in synchronous and asynchronous learning networks, *The 35th Hawaii International Conference on System Sciences*, 1–7 January, Hawaii, 1–9.

Clampitt, P.G. and Downs, C.W. (1993) Employee perceptions of the relationship between communication and productivity: A field study, *Journal of Business Communications*, **30**, 5–28.

Clark, R.A. (1991) *Studying Interpersonal Communication, The Research Experience*. London: Sage.

Cline, R.J.W. (1994) Groupthink and the Watergate cover-up: The illusion of unanimity, in: L.R. Frey (ed.) *Group Communication in Context: Studies of Natural Groups*. New Jersey: Lawrence Erlbaum Associates, 199–223.

Collaros, P.A. and Anderson, L.R. (1969) Effect of perceived expertness upon creativity of members of brain-storming groups, *Journal of Applied Psychology*, **53** (2), pt. 1, 159–163.

Cook, M. and Pasquire, C. (2001) The relationship between the management of design projects and improved performance in the building services sector of the construction industry, *Association of Researchers in Construction Management, 17th Annual Conference*, 5–7 September, University of Salford, 723–730.

Cox, A. and Ireland, P. (2002) Managing construction supply chains: The common sense approach, *Engineering, Construction and Architectural Management*, **5/6**, 409–418.

Cox, A. and Townsend, T. (1998) *Strategic Procurement in Construction: Towards Better Practice in the Management of Construction Supply Chains*. London: Thomas Telford Publishing.

Craig, S. and Jassim, H. (1995) *People and Project Management for IT*. Maidenhead: McGraw-Hill Book Company Europe.

Creyer, E.H. (1997) The influence of firm behaviour on purchase intention, *Journal of Consumer Marketing*, **14** (6), 421–432.

Cummings, D. (2000) Running effective meetings, *Construction Manager*, December, 37.

Cyert, R. and March. J. (1963) *A Behavioural Theory of the Firm*. Englewood Cliffs, New Jersey: Prentice-Hall.

Dainty, A., Moore, D. and Murray, M. (2006) *Communication in Construction: Theory and Practice*. London: Taylor and Francis.

Daly, J.A., McCroskey, J.C., Ayres, J., Hopf, T. and Ayres, D.M. (1997) *Avoiding Communication: Shyness, Reticence, and Communication Apprehension*, 2nd edn. Cresskil, New Jersey: Hampton Press.

Dance, F.E.X. and Larson, C.E. (1972) *Speech Communication: Concepts and Behaviour*. New York: Holt, Rinehart and Winston.

De Grada, E., Kruglanski, A.W., Mannetti, L. and Pierro, A. (1999) Motivation cognition and group interaction: Need for closure affects the contents and processes of collective negotiations, *Journal of Experimental Social Psychology*, **35**, 346–365.

den Otter, A. (2005) *Design Team Communication and Performance Using a Project Website*. PhD thesis (Bouwstenen 98), Technical University of Eindhoven.

Derbyshire, M.E. (1972) *Balancing Objectives in Building*. PhD thesis, University of Lancaster.

Deutsch, M. (1949) An experimental study of the effects of co-operation and competition upon group processes, *Human Relations*, **2**, 199–232.

Downing, M. (1996) It's all about better communication, *Building Products*, October, 12.

Drew, P. and Heritage, J. (1992) *Talk at Work: Interaction in Institutional Settings*. Cambridge: Cambridge University Press.

Duncan, S. (1972) Some signals and rules for taking speaking turns in conversation, *Journal of Personality and Social Psychology*, **23**, August, 283–292.

Duncan, S. Jr and Fiske, D.W. (1977) *Face to Face Interaction: Research, Methods and Theory*. New Jersey: Lawrence Erlbaum Associates.

Egan, J. (1998) *Rethinking Construction: The Report of the Construction Task Force*, July, London: DETR.

——(2002) *Rethinking Construction: Accelerating Change*. London: Strategic Forum for Construction.

Egbu, C. and Gorse, C.A. (2002) 71 Teamwork, in: M. Stevens (ed.) *Project Management Pathways – The Essential Handbook for Project and Programme Managers*, High Wycombe: The Association for Project Management, Chapter 71-1 to 71-16.

Egbu, C., Gorse, C.A. and Emmitt, S. (2000a) Innovation and Knowledge Management: Networks and Networking in the Construction Industry, *UICB 2000, Conference Proceedings of CIB W102 Working Group*, Helsinki, Finland.

Egbu, C., Sturges, J. and Gorse, C.A. (2000b) Communication of Knowledge for Innovation within Projects and Across Organisational Boundaries Congress 2000, *APM Association for Project Management, 15th World Congress on Project Management*, London.

Ellis, D.G. and Fisher, B.A. (1994) *Small Group Decision Making: Communication and the Group Process*, 4th edn. New York: McGraw-Hill.

Emmerson Report (1962) *Survey of the Problems Before the Construction Industry*. London: HMSO.

Emmitt, S. (1997) *The Diffusion of Innovations in the Building Industry*. PhD thesis, Faculty of Arts, University of Manchester.

——(1999) *Architectural Management in Practice*. Harlow, UK: Addison-Wesley Longman.

Emmitt, S. and Gorse, C.A. (1996) De-constructing procurement routes opening up for detail, in: S. Emmitt (ed.) *Detail Design in Architecture*. Northampton: BRC, 24–35.

——(1998) Information management the heart of the design agenda, *The International Journal of Architectural Management, Practice and Research*, **14**, 112–120.

——(2003) *Construction Communication*. Oxford: Blackwell Publishing.

Emmitt, S., Gorse, C.A. and Jones, S. (1997) The site agent – A role in transition, *The International Journal of Architectural Management, Practice and Research*, **13**, 180–185.

——(1999) The housing site manager – Changing roles and training needs, *ARCOM Association of Researchers in Construction Management, 15th Annual Conference*, 15–17 September, Liverpool John Moores, 551–558.

Emmitt, S., Sander, D. and Christoffersen, A.K. (2004) Implementing value through lean design, *Proceedings of 12th Annual Conference on Lean Construction* (IGLC). Denmark: Helsingør, 361–374.

Fangel, M. (ed.) (2005) *Competencies in Project Management: National Competence Baseline for Scandinavia*, Association of Danish Project Management. Denmark: Hillerød.

Farmer, S.M. and Roth, J. (1998) Conflict-handling behaviour in work groups, *Small Group Research*, **29** (6), 669–698.

Fellows, R. and Liu, A. (1999) *Research Methods for Construction*. Oxford: Blackwell Science.

Fenn, P. (1992) Managing the contractual relationship: Privatisation and project management. *Association of Researchers in Construction Management, 8th Annual Conference*, 18–20 September, Douglas Isle of Man, 81–87.

Fenn, P. and Gameson, R. (eds) (1992) Construction conflict management and resolution, *Proceedings of the First International Construction Management Conference*. London: E & FN Spon.

Fenn, P., Lowe, D. and Speck, C. (1997) Conflict and dispute in construction, *Construction Management and Economics*, **15**, 513–518.

Fielding, N.G. and Fielding, J.L. (1986) *Linking Data*. London: Sage Publications.

Fiske, J. (1990) *Introduction to Communication Studies*, 2nd edn. London: Routledge.

Fledman, D.C. (1981) The multiple socialization of organization members, *Academic Management Review*, **6**, 309–318.

Fledman, D.C. (1984) The development and enforcement of group norms, *Academy of Management Review*, **9**, 47–53.

Folger, J.P. and Poole, M.S. (1984) *Working Through Conflict: A Communication Perspective*. Glenview, Illinois: Scott, Foresman.

Franks, J. (1991) *Building Contract Administration and Practice*. CIOB, London: Batsford Ltd.

——(1992) Construction conflict management: The role of education and training, in: P. Fenn and R. Gameson (Eds) *Proceedings of the First International Construction Management Conference*, UMIST. London: E & FN Spon, 406–415.

Frey, L.R. (1994) *Group Communication in Context: Studies of Natural Groups.* New Jersey: Lawrence Erlbaum Associates.

——(1999) *The Handbook of Group Communication Theory and Research.* London: Sage.

Frost, P.J. (1987) Power, politics, and influence, in: F.M. Jablin, L.L. Putnam, K.H. Roberts and L.W. Porter (eds) *Handbook of Organizational Communication.* Beverly Hill, California: Sage, 503–548.

Frye, R.L. (1966) The effect of orientation and feedback of success and effectiveness on the attractiveness and esteem of the group, *Journal of Social Psychology*, **70**, December, 205–211.

Fryer, B., Egbu, C., Ellis, R. and Gorse, C.A. (2004) *The Practice of Construction Management*, 3rd edn. Oxford: Blackwell Science.

Furnham, A. (1986) Situational determinants of intergroup communication, in: W.B. Gudykunst (ed.) *Intergroup Communication: The Social Psychology of Language*, **5**. Baltimore, Maryland: Edward Arnold Publishers.

Gallagher, K., Rose, E., McClelland, B., Reynolds, J. and Tombs, S. (1997) *People in Organisations, An Active Learning Approach.* Oxford: Blackwell.

Gameson, R.N. (1992) *An Investigation into the Interaction Between Potential Building Clients and Construction Professionals.* PhD thesis, Department of Construction Management and Engineering, University of Reading.

Gardiner, P.D. and Simmons, J.E.L. (1992) Analysis of conflict and change in construction projects, *Construction Management and Economics*, **10**, 459–478.

Gašparík, J. (1999) The roles of architect, constructor and owner in a process of quality management, *The International Journal of Architectural Management*, **15**, 6–15.

Gemmill, G.R. and Thamhain, H.J. (1974) Influencing Styles of Project Managers, *Business Management Journal*, **17** (2), 216–224.

Gibb, J.R. (1961) Defensive communication, *Journal of Communication*, 141–148.

Giles, H. (1986) General preface, in: W.B. Gudykunst (ed.) *Intergroup Communication: The Social Psychology of Language*, **5**. Baltimore, Maryland: Edward Arnold Publishers.

Godefroy, C.H. and Robert, L. (1998) *The Outstanding Negotiator, How to Develop Your Arguing Power*, 3rd edn. London: Judy Pistkus.

Goffman, E. (1974) *Frame Analysis.* New York: Harper Colophon Books.

Goleman, D. (1996) *Emotional Intelligence.* London: Bloomsbury.

Gorse, C.A. (2002) *Effective Interpersonal Communication and Group Interaction During Construction Management and Design Team Meetings.* PhD thesis, University of Leicester.

——(2003) Conflict and conflict management in construction, in: D. Greenwood (ed.) *Proceedings of the Association of Researchers in Construction Management, 19th Annual Conference*, 3–5 September, Brighton University, 173–182.

Gorse, C.A. and Archer, I. (1998) Integration of information technology a human perspective, in S. Emmitt (ed.) *The Product Champions: Proceedings of the 2nd International Conference on Detail Design in Architecture*, November, Leeds Metropolitan University.

Gorse, C.A. and Emmitt, S. (1998a) Information exchange between the architect and the contractor, in: Emmitt, S. (ed.) *The Product Champions: Proceedings of the 2nd International Conference on Detail Design in Architecture*, **2**, November, 182–199.

——(1998b) Communication medium the architect and contractor, *The International Journal of Architectural Management, Practice and Research*, **14**, 140–153.

——(2003) Investigating interpersonal communication during construction progress meetings: Challenges and opportunities, *Engineering, Construction and Architectural Management*, **10** (4), 234–244.

——(2004) Management and design team communication, practice section, in: R. Ellis and M. Bell (eds) *Proceedings of COBRA 2004 The International Construction Conference, Responding to Change*, 7–8 September, RICS.

——(2005a) Group interaction research methods, in: D. Greenwood (ed.) *Proceedings of the Association of Researchers in Construction Management, 21st Annual Conference*, 7–9 September, ARCOM, London.

——(2005b) Small group interaction research methods, *Designing Value: Proceedings of the International Conference of Architectural Management. New directions in Architectural Management, CIB W096*, 2–4 November, Technical University of Denmark, Denmark.

Gorse, C.A. and Whitehead, P. (2002) The teaching, learning, experiencing and reflecting on the decision-making process, in: F. Khosrowshahi (ed.) *DMinUCE the 3rd International Conference on Decision Making in Urban and Civil Engineering*, 6–8 November, London (CDRom).

Gorse, C.A., Emmitt, S. and Lowis, M. (1999a) Problem solving and appropriate communication medium, in: W. Hughes (ed.) *Association of Researchers in Construction Management, 15th Annual Conference*, 15–17 September, Liverpool John Moores University, 511–518.

Gorse, C.A., Sturges, J. and Emmitt, S. (1999b) Avoiding Communication, in: P. Nicholson (ed.) *Proceedings of the 15th Meeting of CIB W96 University of Venice*, 19–21 March, Nottingham, UK, SAAM, **15**, 110–118.

Gorse, C.A., Emmitt, S., Lowis, M. and Howarth, A. (2000a) A critical examination of methodologies for studying human communication in the construction industry. *ARCOM Association of Researchers in Construction Management, 16th Annual Conference*, 6–8 September, Glasgow Caledonian University, 31–39 (ISBN 0 9534161 4 3 (2 vols)).

——(2000b) Interaction analysis during management and design team meetings. *ARCOM Association of Researchers in Construction Management, 16th Annual Conference*, 6–8 September, Glasgow Caledonian University, ARCOM Reading, 763–771 (ISBN 0 9534161 4 3 (2 vols)).

——(2000c) A methodology for research of construction communication, in: S. Emmitt (ed.), *Detail Design in Architecture*, 12–13 September, Brighton University, **3**, 41–50.

——(2000d) Models of group and interpersonal interaction during management and design team meetings. in: S. Emmitt (ed.), *Detail Design in Architecture*, 12–13 September, Brighton University, **3**, 50–60.

——(2001) Project performance and management and design team communication, in: A. Akintoye (ed.) *Proceedings of the Association of Researchers in Construction Management, 17th Annual Conference*, 5–7 September, University of Salford.

——(2002) Interaction characteristics of successful contractor's representatives, in: D. Greenwood (ed.) *Proceedings of the Association of Researchers in Construction Management, 18th Annual Conference*, 2–4 September, Northumbria University, 187–188.

Gottschalk, L.A. (1974) Quantification and psychological indicators of emotions: The content analysis of speech and other objective measures of psychological measures, *International Journal of Psychiatry in Medicine*, **5**, 587–610.

Gouran, D.S. (1999) Communication in groups, the emergence and evolution of a field of study, in: L.R. Frey (ed.) *The Handbook of Group Communication Theory and Practice*. London: Sage, 3–36.

Gouran, D.S. and Hirokawa, R.Y. (1986) Counteractive functions of communication in effective group decision making, in: R.Y. Hirokawa and M.S. Poole (eds) *Communication and Group Decision-making*. Beverly Hills, California: Sage, 81–90.

——(1996) Function theory and communication in decision-making and problem-solving groups: An expanded view, in: R.Y. Hirokawa and M.S. Poole (eds) *Communication and Group Decision Making*. 2nd edn. Thousand Oaks, CA: Sage, 55–80.

Gray, C. and Hughes, W. (2001) *Building Design Management*. Oxford: Butterworth-Heinemann.

Gruenfeld, L.W. and Lin, T.R. (1984) Social behaviour of field independents and dependents in an organic group, *Human Relations*, **37**, 721–741.

Gudykunst, W.B. (1986) *Intergroup Communication. The Social Psychology of Language*. Baltimore, Maryland: Edward Arnold Publishers, **5**.

Guevara, J.M. and Boyer, L.T. (1981) Communication problems within construction, *Journal of the Construction Division*, Proceedings of the ASCE, **107** (CO4), December, 551–557.

Gupta, S. and Case, T.L. (1999) Managers' outward influence tactics and their consequences, *Leadership and Organisational Development Journal*, **20** (6), 300–308.

Gushgari, S.K., Francis, P.A. and Saklou, J.H. (1997) Skills critical to long-term profitability of engineering firms, *Journal of Management in Engineering*, March, 46.

Gutzmer, W.E. and Hill, W.F. (1973) Evaluation of the effectiveness of the learning through discussion method, *Small Group Behaviour*, **4** (1), February, 5–34.

Guzzo, R.A. (1995) Introduction: At the intersection of team effectiveness and decision making, in: R.A. Guzzo and E. Salas (eds) *Team Effectiveness and Decision Making in Organizations*. San Francisco: Jossey-Bass, 1–8.

Hackman, J.R. (1992) Group influences on individuals in organizations, in: M.D. Dunnette and L.M. Hough (eds) *Handbook of Industrial and Organizational Psychology*, 2nd edn. Palo Alto, California: Consulting Psychologists Press.

Hackman, J.R. and Morris, C.G. (1975) Group tasks, group interaction process, and group performance effectiveness: A review of proposed integration, in: L. Berkowitz (ed.) *Advances in Experimental Social Psychology*. New York: Academic Press, **8**, 45–99.

Hackman, J.R. and Vidmar, N. (1970) Effects of size and task type on group and member reactions, *Sociometry*, **33** (1), March, 37–54.

Hakin, C. (1987) *Research Design: Strategies and Choices in the Design of Social Research*. London: Allen and Unwin.

Hall, M. (2001) 'Root' cause analysis: A tool for closer supply chain integration in construction, *Association of Researchers in Construction Management, 17th Annual Conference*, 5–7 September, University of Salford, 929–938.

Halpin, N.R. (1990) *Consistency of Personality Traits and Verbal Behaviour in Leaderless Small Group Discussions*. PhD thesis, Department of Psychology, University of Durham.

Hancock, R.D. and Sorrentino, R.M. (1980) The effects of expected future interactions and prior group support on the conformity process. *Journal of Experimental Social Psychology*, **16** (3), 261–269.

Hanlon, M.D. (1980) Observational methods in organisational assessment, in: D.D. Lawer (III), D.A. Nadler and C. Cammann (eds) *Organisational Assessment: Perspectives on the Measurement of Organisational Behaviour and the Quality of Working Life, Series on Organisational Assessment and Change*. New York: John Wiley & Sons, 349–381.

Hare, A.P. (1976) *Handbook of Small Group Research*, 2nd edn. New York: Free Press.

Hargie, O. (1992) Communication: Beyond the crossroads, *Monograph*. Jordanstown: University of Ulster.

Hargie, O.D.W., Dicksos, D. and Tourish, D. (1999) *Communication in Management*. Hampshire England, Gower.

Hartley, P. (1997) *Group Communication*. London: Routledge.

Harvey, R.C. and Ashworth, A. (1997) *The Construction Industry of Great Britain*, 2nd edn. Oxford: Butterworth-Heinemann.

Harwood, T. (2002) Business negotiations in the context of strategic relationship development, *Marketing Intelligence and Planning Journal*, **20** (6), 336–348.

Haslett, B.B. and Ruebush, J. (1999) What differences do individual differences in groups make? The effects of individuals, culture and group composition, in: F.R. Frey (ed.) *The Handbook of Group Communication Theory and Research*. London: Sage, 115–139.

Hastings, I. (1998) The virtual project team, *Project Manager Today*, July, 26–29.

Hawkins, K. and Power, C.B. (1999) Gender differences in question asked during small decision-making group discussions, *Small Group Research*, **30** (2), 235–256.

Hayes, J. (1991) *Interpersonal Skill, Goal-Directed Behaviour at Work*. London: HarperCollins Academic.

Heinicke, C. and Bales, R.F. (1953) Developmental trends in the structure of small groups, *Sociometry*, **16**, 7–38.

Higgin, G. and Jessop, N. (1965) *Communication in the Building Industry: The Report of a Pilot Study*. The Tavistock Institute of Human Relations. London: Tavistock Publications.

Hill, C.J. (1995) Communication on construction sites, *Association of Researchers in Construction Management, 11th Annual Conference*, 18–20 September, University of York, 232–240.

Hiltz, S.R. and Turoff, M. (1982) *The Network Nation: Human Communication via Computer*. London: Addison-Wesley.

——(1993) *The Network Nation: Human Communication via Computer*. Revised edition, MIT Press.

Hirokawa, R.Y. and Poole, M.S. (1996) *Communication and Group Decision Making*, 3rd edn. London: Sage.

Hirokawa, R.Y. and Salazar, A.J. (1999) Task-group communication and decision-making performance, in: L.R. Frey (ed.) *The Handbook of Group Communication Theory and Research*. London: Sage, 167–191.

Hirokawa, R.Y. and Scheerhorn, D.R. (1986) Communication in faulty group decision-making, in: R.Y. Hirokawa and M.S. Poole (eds) *Communication and Group Decision Making*. Beverly Hills: Sage, 63–80.

Hodgetts, R.M. (1968) Leadership techniques in the project organization, *Business Management Journal*, 11, 15–25.

Hoffman, J. and Arsenian, J. (1965) An example of some models applied to group structures and process, *International Journal of Group Psychotherapy*, **15**, 131–153.

Hollingshead, A.B. (1996) The rank order effect in group decision making, *Organizational Behaviour and Human Decision Processes*, **68** (3), December, 181–193.

——(1998) Communication, learning and retrieval in transactive memory systems, *Journal of Experimental Social Psychology*, **34**, 423–442.

Hosking, D. and Haslam, P. (1997) Managing to relate: Organizing as a social process, *Career Development International*, **2** (2), 85–89.

Hughes, W.P. (1994) The PhD in construction management, *Association of Researchers in Construction Management, 10th Annual Conference*, 14–16 September, Loughborough University of Technology, 76–87.

Hugill, D. (1998) Illuminating a psychological theory (in a construction management context), *Association of Researchers in Construction Management, 14th Annual Conference*, 9–11 September, University of Reading, 22–30.

——(1999) Negotiating access: Presenting a credible project, in: W. Hughes (ed.), Proceedings 15th *Annual ARCOM Conference*, 15–17 September, Liverpool John Moores University, 53–63.

——(2000) Management as an accomplishment of project team meetings in construction, in: A. Akintoye (ed.) *Proceedings 16th Annual ARCOM Conference*, 6–8 September, Glasgow Caledonian University, 755–762.

——(2001) *An Examination of Project Management Team Meetings in Railway Construction*. PhD thesis, Faculty of Economics, Social Studies and Law, University of Manchester.

Huseman, R.C. (1977) Interpersonal conflict in the modern organisation, in: R.C. Huseman, C.M. Logue and D.L. Freshley (eds) *Readings in Interpersonal and Organisational Communication*, 3rd edn. London: Allyn and Bacon, 222–233.

Huseman, R.C., Logue, C.M. and Freshley, D.L. (1977) *Readings in Interpersonal and Organisational Communication*, 3rd edn. London: Allyn and Bacon.

Jablin, F.M. (1982) Organizational communication: An assimilation approach, in: M.E. Roloff and C.R. Berger (eds) *Social Cognition and Communication*. Beverly Hills, California: Sage, 255–286.

——(1984) Assimilation new members into organizations, in: R.N. Bostrom (ed.) *Communication Yearbook*, **8**. Beverly Hills, California: Sage.

Jackson, J. (1965) Social stratification, social norms, and roles, in: I.D. Steiner and M. Fishbein (eds) *Current Studies in Social Psychology*. New York: Holt Rinehart and Wiston, 301–309.

Jacobs, A., Jacobs, M., Cavior, N. and Burke, J. (1974) Anonymous feedback: Credibility and desirability of structured emotional and behavioral feedback delivered in groups, *Journal of Counselling Psychology*, **21** (2), March, 106–111.

Jaffe, J., Lee, Y., Huang, L. and Oshagan, H. (1995) Gender, pseudonyms and CMC: Masking identities and baring souls, *International Communication Association*. Albuquerque: New Mexico.

Janis, I.L. (1982) *Groupthink: Psychological Studies of Policy Decisions and Fiascos*, 2nd edn. Boston: Houghton Mifflin.

Jankowicz, A.D. (2005) *Business Research Projects*, 4th edn. London: Thomson Learning.

Jarboe, S. (1988) A comparison of input-output, process-output, and input-process-output models of small group problem solving effectiveness, *Communication Monographs*, **55**, 121–142.

Katz, D. and Kahn, R.L. (1978) *The Social Psychology of Organizations*, 2nd edn. New York: John Wiley.

Katzell, R.A., Miller, C.E., Rotter, N.G. and Venet, T.G. (1970) Effects of leadership and other inputs on group process and outputs. *Journal of Social Psychology*, **80**, 157–169.

Kelly, C. (2003) *Increasing Competition and Improving Long-Term Capacity Planning in the Government Market Place* (The Kelly Report), Office of Government Commerce, London.

Ketrow, S.M. (1999) Nonverbal aspects of group communication, in: L.R. Frey (ed.) *The Handbook of Group Communication Theory and Research*. London: Sage, 251–287.

Keyton, J. (1999) Relational communication in groups, in: L.R. Frey (ed.) *The Handbook of Group Communication Theory and Research*. London: Sage, 192–221.

## 286    Bibliography

——(2000) Introduction: The relational side of groups, *Small Group Research*, **31** (4), 387–394.

Kirscht, J.D., Lodahl, T.M. and Haire, M. (1959) Some factors in the selection of leaders by members of small groups, *Journal of Abnormal and Social Psychology*, **58**, 406–408.

Klob, D. (1992) *Hidden Conflict in Organisations*. London: Sage Publications.

Kreiner, K. (1976) *The Site Organisation: A Study of Social Relationships on Construction Sites*. PhD thesis, Technical University of Denmark, Denmark.

Kreps, G.L. (1989) *Organizational Communication Theory and Practice*, 2nd edn. New York: Longman.

Krippendorff, K. (1980) *Content Analysis: An Introduction to its Methodology*. London: Sage Publications.

Kristiansen, K., Emmitt, S. and Bonke, S. (2005) Changes in the Danish construction sector: The need for a new focus, *Engineering Construction and Architectural Management*, **12** (5), September/October, 502–511.

Kruglanski, A.W. (1990) Motivations for judging and knowing: Implications for causal attribution, in: E.T. Higgins and R.M. Sorrentinon (eds) *The Handbook of Motivation and Cognition*. New York: Guilford, **2**, 333–368.

Kwakye, A. (1997) *A Construction Project Administration in Practice*, CIOB. Harlow: Addision Wesley Longman.

Lamm, H. and Trommsdorff, G. (1973) Group versus individual performance tasks requiring ideational proficiency: A review, *European Journal of Social Psychology*, **3**, 361–388.

Landsberger, H. (1955) Interaction process analysis of professional behaviour: A study of labour mediators in 12 labour-management disputes, *American Sociological Review*, **20**, 566–575.

Langford, D., Hancock, M.R., Fellows, R. and Gale, A.W. (1995) *Human Resources Management in Construction*. Harlow: Longman Scientific & Technical.

Larson, C.W. and LaFasto, F.M.J. (1989) *Teamwork: What Must Go Right/What can Go Wrong*. Beverly Hills, California: Sage.

Larsson, B. (1992) Adoption av ny produktionsteknik på byggarbetsplatsen (Adoption of new building technologies on the building site). PhD thesis, Report 30, Chalmers University of Technology, Gothenberg, Sweden.

Latham, M. (1993) *Trust and Money: Interim Report of the Joint Government Industry Review of Procurement and Contractual Arrangements in the United Kingdom Construction Industry*. London: HMSO.

——(1994) *Constructing the Team*, Final Report. London: HMSO.

Lavers, A.P. (1992) Communication and clarification between designer and client: Good practice and legal obligation, in: M.P. Nicholson (ed.) *Architectural Management*. London: E & FN Spon, 23–28.

Lawler, E.E. (III) (1980) Adaptive experiments, in: D.D. Lawer (III), D.A. Nadler, and C. Cammann (eds) *Organisational Assessment: Perspectives on the Measurement of organisational Behaviour and the quality of Working Life*, Series on organisational assessment and change. New York: John Wiley & Sons, 101–113.

Lawson, B.R. (1970) Open and closed ended problem solving in architectural design, in: Honickman. B (ed.) A.P. (70) *Proceedings of the Architectural Psychology Conference*, Kingston Polytechnic. London: RIBA Publications, 87–91.

——(1972) *Problem Solving in Architectural Design*. Unpublished PhD thesis, University of Birmingham.

Lederman, W. (1984a) *Handbook of Applicable Mathematics, VI, Part A: Statistics.* Chichester: John Wiley & Sons.

——(1984b) *Handbook of Applicable Mathematics, VI, Part B: Statistics.* Chichester: John Wiley & Sons.

LeDoux, J. (1998) *The Emotional Brain.* New York: *Phonix.*

Lee, F. (1997) When the going gets tough, do the tough ask for help? Help seeking and power motivation in organizations, *Organizational Behaviour and Human Decision Processes,* **72** (3), December, 336–363.

Lesch, C.L. (1994) Observing theory in practice: Sustaining consciousness in a coven, in: L.R. Frey (ed.) *Group Communication in Context: Studies of Natural Groups.* Hove, UK: Lawrence Erlbaum Associates, 57–82.

Lewis, J. and Cheetham, D.W. (1993) The historical roots of current problems in building procurement, *Association of Researchers in Construction Management, 9th Annual Conference,* 14–16 September, Oxford University, 50–61.

Lieberman, M., Lakin, M. and Whitaker, D. (1969) Problems and potential of psycho-analytic and group theories for group psychotherapy, *International Journal of Group Psychotherapy,* **19**, 131–141.

Littlepage, G.E. and Silbiger, H. (1992) Recognition of expertise in decision-making groups: Effects of group size and participation patterns, *Small Group Research,* **22**, 344–355.

Lonetto, R. and Williams, D. (1974) Personality, behavioural and output variables in a small group task situation: An examination of consensual leader and non-leader differences, *Canadian Journal of Behavioural Science,* **6**, 58–74.

Loosemore, M. (1992) Managing the construction process through a framework of decisions, in: M.P. Nicholson (ed.) *Architectural Management,* London: Spons, 90–103.

——(1993) Communication network analysis: A powerful new tool for construction management research, *Association of Researchers in Construction Management, 9th Annual Conference,* 14–16 September, Oxford University, 411–423.

——(1994) Problem behaviour, *Construction Management and Economics,* **12**, 511–520.

——(1995) Construction crisis: Investigating links between communication structure and behaviour, *Association of Researchers in Construction Management, 11th Annual Conference,* 18–20 September, University of York, 241–250.

——(1996a) *Crisis Management in Building Projects: A Longitudinal Investigation of Communication Behaviour and Patterns Within a Grounded Framework.* PhD thesis, Department of Construction Management and Engineering, University of Reading.

——(1996b) The problems of data collection during construction crises, *Association of Researchers in Construction Management, 12th Annual Conference,* 11–13 September, Sheffield Hallam University, 488–495.

——(1998a) The influence of communication structure upon crisis management efficiency, *Construction Management and Economics,* **16**, 661–671.

——(1998b) The methodological challenges posed by the confrontational nature of the construction industry, *Engineering, Construction and Architectural Management,* **3** (5), September, 285–294.

——(1999) Responsibility, power and construction conflict, *Construction Management and Economics,* **17**, 699–709.

Loosemore, M. and Chin Chin Tan (2000) Occupational stereotypes in the construction industry, *Construction Management and Economics,* **18**, 559–566.

Loosemore, M., Nguyen, B.T. and Denis, N. (2000) An investigation into the merits of encouraging conflict in the construction industry, *Construction Management and Economics,* **18**, 447–456.

Luft, J. (1984) *Group Process: An Introduction to Group Dynamics*. Palo Alto, California: Mayfield.

Mabry, E.A. (1985) The effects of gender composition and task structure on small group interaction, *Small Group Behavior*, **16** (1), February, 75–96.

——(1989) Some theoretical implications of female and male interaction in small unstructured groups, *Small Group Behavior*, **20** (4), November, 536–550.

——(1999) The systems metaphor in group communication, in: L.R. Frey (ed.) *The Handbook of Group Communication Theory and Research*. London: Sage, 71–91.

Mackinder, M. (1980) *The Selection and Specification of Building Materials and Components*, Research paper 17, University of York Institute for Advanced Architectural Studies.

Mackinder, M. and Marvin, H. (1982) *Design Decision-making in Architectural Practice*, Institution for Advance Architectural Studies, University of York, Research paper 19, IAAS, New York.

Macmillan, S. (2001) The measure of a team, *Construction Manager*, July/August, 25.

Manheim, H.L. (1963) Experimental demonstration of relationship between group characteristics and patterns of intergroup interaction, *Pacific Sociological Review*, **5**, Spring, 25–29.

Mann, R.D. (1961) Dimensions of individual performance in small groups under task and social-emotional conditions, *Journal of Abnormal and Social Psychology*, **62** (3), May, 674–682.

Masnikosa, V.P. (1999) On some obstacles in communication and transfer of knowledge, *Kybernetes*, **28** (5), 575–584.

McCann, D. (1993) *How to Influence Others at Work: Psychoverbal Communication for Managers*, 2nd edn. London: Butterworth-Heinemann.

McCroskey, J.C. (1977) Oral communication apprehension: A summary of recent theory and research, *Human Communication Research*, **4**, 78–96.

——(1997) Willingness to communication, communication apprehension, and self-perceived communication competence: Conceptualizations and perspectives, in: J.A. Daly, J.C. McCroskey, J. Ayres, T. Hopf and D.M. Ayres (eds) *Avoiding Communication: Shyness Reticence, and Communication Apprehension*. New Jersey: Hampton Press, 75–108.

McCroskey, J.C. and Richmond, V.P. (1990) Willingness to communicate: A cognitive view, *Journal of Social Behavior and Personality*, **5**, 19–37.

McGrath, J.E. (1984) *Groups; Interaction and Performance*. Englewood Cliffs, New Jersey: Prentice-Hall.

——(1997) Small group research, that once and future field: An interpretation of the past with an eye to the future, *Group Dynamics: Theory, Research and Practice*, **1**, 7–27.

Meister, J. and Reinsch, N.L. Jr (1978) Communication training needs in manufacturing firms, *Communication Education*, **27**.

Melvin, T. (1979) *Practical Psychology in Construction Management*. New York: Van Nostrand Reinhold company.

Meyers, R.A. and Brashers, D.E. (1999) Influence processes in group interaction, in: L.R. Frey (ed.) *The Handbook of Group Communication Theory and Research*. London: Sage, 288–312.

Mills, T.M. (1967) *The Sociology of Small Groups*. New Jersey: Prentice-Hall.

Moore, D. (2002) *Project Management: Designing Effective Organisational Structures in Construction*. Oxford: Blackwell Publishing.

Morgan, B.B., and Bowers, C.A. (1995) Teamwork stress: Implications for team decision making, in: R.A. Guzzo and E. Salas (eds) *Team Effectiveness and Decision Making in Organizations*. San Francisco: Jossey-Bass, 262–290.

Morley, I.E. (1984) *Bargaining and Negotiating*. London: Macmillan Publishers.

Morris, C.G. (1966) Task effects on group interaction, *Journal of Personality and Social Psychology*, **4** (5), November, 545–554.

Morrison, E.W. (1993) Newcomer information seeking: Exploring types, models, sources and outcomes, *Academy of Management Journal*, **36** (3), 556–589.

Mostyn, B. (1985) The content analysis of qualitative research data, in: M. Brenner, J. Brown and D. Canter (eds) *The Research Interviews Uses and Approaches*. London: Academic Press, 115–145.

Mullen, B., Salas, E. and Driskell, J.E. (1989) Salience, motivation and artefact as contributions to the relation between participation rate and leadership, *Journal of Experimental Social Psychology*, **25**, 545–559.

Murray, J.S. (1986) Understanding competing Theories of Negotiation, *Negotiation Journal*, **2**, 179–186.

Mutti, C.N. and Hughes, W. (2001) Contemporary organizational theory in the management of construction projects, *Association of Researchers in Construction Management, 17th Annual Conference*, 5–7 September, University of Salford, 455–465.

Naoum, S.G. and Mustapha, F.H. (1995) Relationship between the building team, procurement methods and project performance. *Journal of Construction Procurement*, **1** (1), 50–63.

NEDO (1985) *Thinking About Building*. London: HMSO.

Needham, M.J. (1998) *Arbitration for the Construction Industry*, Annual guest lecture series, at Leeds Metropolitan University, School of the Built Environment. An unpublished paper delivered on 25 March.

Nicholson, P. (1997) Design and build is not enough. *Construction Manager*, **3** (8), October, 55.

Norusis, M.J. (1998) *SPSS 8.0, Guide to Data Analysis*. New Jersey: Prentice-Hall.

Olmsted, S. (1979) *The Small Group*. London: Random House.

O'Mahony, T. (1977) *An Appraisal of Factors Affecting Communications Within a Building Design Team*. Unpublished MSc thesis, Department of Civil Engineering, University of Salford.

Osgood, C.E., Saporta, S. and Nunnally, J.C. (1959) Evaluative assertion analysis, *Litera*, **3**, 47–102.

Oxman, R. (1995) Multi-agent collaboration in integrated information systems. In: P. Brandon and M. Bretts (eds) *Integrated Construction Information*. London: E & FN Spon, 317–325.

Paciocco, T. (2000) The practice of architecture, the survival of the firm, *The International Journal of Architectural Management*, CIB W096, **15**, 94–104.

Patchen, M. (1993) Reciprocity of coercion and co-operation, in: R.B. Felson and J.T. Tedeschi (eds) *Aggression and Violence: Social Interactionist Perspectives*. Washington: American Psychological Association, 119–144.

Patton, M.Q. (1990) *Qualitative Evaluation and Research Methods*, 2nd edn. Thousand Oaks: Sage.

Pavitt, C. (1999) Theorizing about the group communication-leadership relationship. Input-process-output and functional models, in: L.R. Frey (ed.) *The Handbook of Group Communication Theory and Research*. London: Spon, 313–334.

Pena, J. (2004) *An Interaction Process Analysis of Text Based Communication in an Online Multiplayer Videogame*, Faculty of the Graduate School, Cornell University, Ithica, **51**.

Philipsen, G. and Albrecht, T.L. (1997) *Developing Communication Theories*. New York: State University Press.

Phillips Report (1950) *Report of the Working Party on the Building Industry*. London: HMSO.

Pietroforte, R. (1992) *Communication and Information in the Building Delivery Process*. PhD thesis, Department of Civil Engineering, Massachusetts Institute of Technology, Cambridge, Massachusetts.

——(1997) Communication and governance in the building process, *Construction Management and Economics*, **15**, 71–82.

Pondy, L. (1967) Organisational conflict, concepts and models, *Administrative Quarterly*, **12**, September, 299–306.

Poole, M.S. (1981) Decision development in small groups, I: A comparison of two models, *Communication Monographs*, **48**, 1–24.

——(1983a) Decision development in small groups, II: A study of multiple sequences in decisions making, *Communication Monographs*, **50**, 206–232.

——(1983b) Decision development in small groups, III: A multiple sequence model of decision development, *Communication Monographs*, **50**, 321–341.

——(1999) Group communication theory, in: L.R. Frey (ed.) *The Handbook of Group Communication Theory and Research*. London: Sage, 37–70.

Poole, M.S. and Baldwin, C.L. (1996) Developmental processes in group decision making, in: R.Y. Hirokawa and M.S. Poole (eds) *Communication and Group Decision Making*, 2nd edn. Thousand Oaks, California: Sage, 215–241.

Poole, M.S. and Hirokawa, R.Y. (1996) Communication and group decision making, in: Hirokawa, R.Y. and Poole, M.S. (eds) *Communication and Group Decision Making*. London: Sage, 3–18.

Poole, M.S., Keyton, J. and Frey, L.R. (1999) Group communication methodology: Issues and considerations, in: L.R. Frey (ed.) *The Handbook of Group Communication Theory and Research*. London: Sage, 92–112.

Potter, J. and Wetherell, M. (1987) *Discourse and Social Psychology: Beyond Attitudes and Behaviour*, London: Sage.

Preece, C. and Tarawneh, S. (1997) Why are design and build clients unhappy? *Construction Manager*, **3** (7), September, 24–25.

Price, S. (1996) *Communication Studies*. Harlow: Longman.

Pruitt, D.G. (1981) *Negotiating Behaviour*. New York: Academic Press Publishers.

Pruitt, D.G., Mikolic, J.M., Peirce, R.S. and Keating, M. (1993) Aggression as a struggle tactic in social conflict, in: R.B. Felson and J.T. Tedeschi (eds) *Aggression and Violence: Social Interactionist Perspectives*. Harvard: American Psychological Association, 99–118.

Psathas, G. (1960) Phase movement and equilibrium tendencies in interaction process analysis in psychotherapy groups, *Sociometry*, **XXIII**, 177–194.

Ragin, C.C. and Becker, H.S. (1992) *What is a Case?: Exploring the Foundations of Social Inquiry*. Cambridge: Cambridge University Press.

Rahim, M.A. (1983) A measure of styles of handling interpersonal conflict, *Academy of Management Journal*, **26** (2), 368–376.

Rajan, S. and Krishnan, V.R. (2002) Impact of influence, power and authoritarianism, *Women in Management Review*, **17** (5), 197–206.

Redmond, K. and Gorse, C.A. (2004) Influencing and persuasion techniques used during initial one-to-one negotiations, in: R. Ellis and M. Bell (eds) *Proceedings of COBRA 2004 the International Construction Conference, Responding to Change*, 7–8 September, RICS.

Reichers, A.E. (1987) An interactionist perspective on newcomer socialization rates, *Academy of Management Review*, **12** (12), 278–287.

Reinard, J. (1997) *Introduction to Communication Research*, 2nd edn. Boston, Massachusetts: McGraw-Hill.

Rim, Y. (1963) Risk taking and need for achievement, *Acta Psycholgica*, **21**, 108–115.

——(1964a) Social attitudes and risk-taking, *Human Relations*, **17**, 259–265.

——(1964b) Personality and group decisions involving risk, *The Psychological Record*, **14**, 37–45.

——(1965) Leadership attitudes and decisions involving risk, *A Journal of Applied Research*, **18**, 423–430.

——(1966) Machiavellianism and decisions involving risk, *British Journal of Social and Clinical Psychology*, **5**, 30–36.

Robson, C. (1993) *Real World Research: A Resource for Social Scientists and Practitioners-Researchers*. Oxford: Blackwell.

Rougvie, A. (1987) *Project Evaluation and Development*. London: Batsford.

Sanders, M., Lewis, P. and Thornhill, A. (2003) *Research Methods for Business Students*, 3rd edn. Harlow: Pearson Education Limited.

Scheerhorn, D. and Geist, P. (1997) Social dynamics in groups, in: L.R. Frey and J.K. Barge (eds) *Managing Group Life: Communicating in Decision-making Groups*. Boston: Houghton Mifflin, 81–103.

Scheidel, T.M. and Crowell, L. (1964) Idea development in small discussion groups, *Quarterly Journal of Speech*, **50**, 140–145.

——(1966) Feedback in group communication, *Quarterly Journal of Speech*, **52**, 273–278.

Schein, E.H. (1987) *Process Consultation: Vol. 2. Lessons for Managers and Consultants*, Reading, Massachusetts: Addison-Wesley.

Schultz, B.C. (1999) Improving group communication performance, an overview of diagnosis and intervention. in: L.R. Frey (ed.) *The Handbook of Group Communication Theory and Research*. London: Sage, 371–394.

Schutz, W. (1973) *Elements of Encounter*, California, BS: Jay Press.

Scott, B. (1988) *Negotiating*. London: Paradigm Publishers.

Senge, P.M. (1990) *The Fifth Discipline: The Art and Practice of the Learning Organization*. New York: Doubleday – Currency.

Seymour, D. and Hill, C. (1993) Implications of research perspective for management policy, *Association of Researchers in Construction Management, 9th Annual Conference*, 14–16 September, Oxford University, 151–123.

Shadish, W.R. (1981) Theoretical observations on applied behavioural science, *Journal of Applied Behavioural Science*, **17** (1), 98–112.

Shapiro, D. and Leiderman, P.H. (1967) Arousal correlates of task role and group setting, *Journal of Personality and Social Psychology*, **5** (1), 103–107.

Shaw, M.E. (1981) *Group Dynamics: The Psychology of Small Group Behaviour*, 3rd edn. New York: McGraw-Hill.

Shaw, M.E., Rothschild, G.H., and Strickland, J.F. (1957) Decision process in communication nets, *Journal of Abnormal Social Psychology*, **54**, 323–330.

Shell, G.R. (1999) *Bargaining, Negotiation Strategies for Reasonable People, for Advantage*. London: Penguin Books Group.

Shepherd, C.R. (1964) *Small Groups: Some Sociological Perspectives*. New York: Chandler Publishing.

Shirazi, A. and Hampson, K. (1998) Project manager competencies in a knowledge based society, *Proceedings of the Association of Researchers in Construction Management, 14th Annual Conference*, 9–11 September, University of Reading, 50–59.

Short, J., Williams, E. and Christie, B. (1976) *The Social Psychology of Telecommunications*. New York: John Wiley.

Silverman, D. (2000) *Doing Qualitative Research: A Practical Handbook*. London: Sage Publications.

——(2001) *Interpreting Qualitative Data*. London: Sage Publications.

Simintiras, A.C. and Cadogan, J.W. (1996) Behaviourism in the study of salesperson – customer interactions, *Journal of Decision Management*, **34** (6), 57–64.

Siminitras, A.C. and Thomas, A.H. (1998) Cross-cultural sales negotiations: A literature review and research propositions, *International Marketing Review*, **15** (1), 10–28.

Simister, S. (1995) Case study methodology for construction management research, *Association of Researchers in Construction Management, 11th Annual Conference*, 18–20 September, University of York, 21–32.

Simon, E.D. (1944) *The Placing and Management of Building Contracts*. London: HMSO.

Smith, J. and Wyatt, R. (1998) Criteria for strategic decision making at the pre briefing stage, *Association of Researchers in Construction Management, 14th Annual Conference*, 9–11 September, University of Reading, 300–309.

Socha, T.J. (1999) Communication in family units: Studying the first 'group', in: L.R. Frey (ed.) *The Handbook of Group Communication Theory and Research*. London: Sage, 475–493.

Socha, T.J. and Socha, D.M. (1994) Children's task-group communication: Did we learn it all in kindergarten?, in: L.R. Frey (ed.) *Group Communication in Context: Studies of Natural Groups*. New Jersey: Lawrence Erlbaum Associates, 227–246.

Sotiriou, D. and Wittmer, D. (2001) Influencing methods of project managers: Perceptions of team members and project managers, *Project Management Journal*, **32** (3), 12–20.

Sperber, D. and Wilson, D. (1986) *Relevance Communication and Cognition*. Oxford: Blackwell.

Stephen, F.F. and Mishler, E.G. (1952) The distribution of participation in small groups: An exponential approximation, *American Sociological Review*, **17**, October, 598–608.

Stohl, C. and Holmes, M.E. (1993) A functional perspective for bona fide groups, in: S.A. Deetz (ed.) *Communication Yearbook*. Newbury Park, California: Sage, **16**, 601–614.

Stone, P., Dunphy, D.C., Smith, M.S. and Ogilvie, D.M. (1966) *The General Inquirer. A Computer Approach to Content Analysis*. Cambridge, MA: MIT Press.

Strauss, A. (1978) *Negotiations; Varieties, Contexts, and Social Order*. USA: Bass Publishers.

Stretton, A. (1981) Construction communications and individual perceptions, *Chartered Builder*, Summer, 51–53.

Streufert, S. and Streufert, S.C. (1969) Effects of conceptual structure, failure and success on attribution of causality and interpersonal attitudes, *Journal of Personality and Social Psychology*, **11** (2), 138–147.

Stroop, J.R. (1932) Is the judgement of the group better than that of the average member of the group? *Journal of Experimental Psychology*, **15**, 550–562.

Sudman, S. (1976) *Applied Sampling*. London: Academic Press.

Sunwolf and Seibold, D.R. (1999) The impact of formal procedures on group processes, members and task outcomes, in: L.R. Frey (ed.) *The Handbook of Group Communication Theory and Research*. London: Sage, 395–431.

Swaffield, L.M. (1998) *Improving Early Cost Advice for Mechanical and Electrical Services*. PhD thesis, Department of Civil Engineering, Loughborough University of Technology.

Swan, W., Cooper, R., McDermott, P. and Wood, G. (2001) Reviews of social network analysis for the IMI trust in construction project, *Association of Researchers in Construction Management, 17th Annual Conference*, 5–7 September, University of Salford, 59–67.

Talland, G.A. (1955) Task and interaction process: Some characteristics of therapeutic group discussion, *Journal of Abnormal and Social Psychology*, 105–109.

Taylor, D.W., Berry, P.C. and Block, C.H. (1958) Does group participation when using brainstroming facilitate or inhibit creative thinking, *Administrative Science Quarterly*, **3**, 23–45.

Taywood Engineering (1997) Poor communication causes defects, *Construction Manager*, Chartered Institute of Building, **3** (5), June, 5.

Thelen, H.A. (1949) Group dynamics in instruction: Principle of least group size, *School Review*, **57**, 139–148.

Thomas, S.R., Tucker, R.L. and Kelley, W.R. (1998) Critical communication variables, *Journal of Construction Engineering and Management*, American Society of Civil Engineers: Construction Division, **124** (4), January/February, 58–66.

Ting Toomey, S. (1986) Interpersonal ties in intergroup communication, in: W.B. Baltimore (ed.) *Intergroup Communication: The Social Psychology of Language*. Grudykunst, Maryland: Edward Arnold, **5**, 114–126.

Trenholm, S. and Jensen, A. (1995) *Interpersonal Communication*, 3rd edn. London: Wadsworth Publishing.

Trujillo, N. (1986) Implications of interpretive approaches for organization communication research and practice, in: L. Thayer (ed.) *Organization-communication: Emerging Perspectives*, Vol. 2: *Communication in Organizations*. Norwood, New Jersey: Ablex, 46–63.

Turk, H. (1961) Instrumental and expressive ratings reconsidered, *Sociometry*, **24**, 76–81.

Viney, L.S. (1983) The assessment of psychological states through content analysis of verbal communications, *Psychological Bulletin*, **94** (3), 542–563.

Wadleigh, P.M. (1997) Contextualizing communication avoidance research, in: J. Daly, J.C. McCroskey, T. Hopf and D. Ayres (eds) *Avoiding Communication*, 2nd edn. Cresskill, New Jersey: Hampton Press, 3–20.

Walker, A. (1996) *Project Management in Construction*, 3rd edn. Oxford: Blackwell Science.

Wallace, W.A. (1987) *The Influence of Design Team Communication Content Upon the Architectural Decision Making Process in the Pre-contract Design Stages*. PhD thesis, Department of Building, Heriot-Watt University.

Wallach, M.S. and Kogan, N. (1959) Sex differences and judgement processes, *Journal of Personality*, **27**, 555–564.

Wasbush, J.B. and Clements, C. (1999) The two faces of leadership, *Career Development International*, **4** (3), 146–148.

Weber, R.J. (1971) Effects of videotape feedback on task group behaviour. *Proceedings of the Annual Convention of the American Psychological Association, 79th Annual Convention*, **6** (2), 499–500.

Weick, K. (1969) *The Social Psychology of Organising*, Reading, Massachusetts: Addison-Wesley.

Wheelan, S.A., McKeage, R.L., Verdi, A.F., Abraham, M., Krasick, C. and Johnston, F. (1994) Communication and developmental patterns in a system of interacting groups. in: L.R. Frey (ed.) *Group Communication in Context: Studies of Natural Groups.* New Jersey: Lawrence Erlbaum Associates, 153–178.

Wild, A. (2001) The Phillips report on building 1950, *Association of Researchers in Construction Management, 17th Annual Conference*, 5–7 September, Salford University, 609–617.

Winter, W.D. and Ferreira, A.J. (1967) Interaction process analysis of family decision-making, Family Process, **6**, 155–172.

Woodcock, B.E. (1979) Characteristics oral and written business communication problems of selected managerial trainees, *Journal of Business Communication*, **16**.

Wyatt, N. (1993) Organizing and relating: Feminist critique of small group communication, in: S.P. Bowen and N. Wyatt (eds) *Transforming Visions: Feminist Critiques in Communication Studies.* Cresskill, New Jersey: Hampton, 21–86.

Yeo, K.T. (1993) System thinking and project management: Time to reunite, *International Journal of Project Management*, **11** (2), 112–120.

Yeomans, D.T. (1970) Assessment of argument and design decisions, in: B. Honickman (ed.) A.P. 70 *Proceedings of the Architectural Psychology Conference at Kingston Polytechnic.* London: RIBA Publications, 2-25-1 to 2-25-8.

Yin, R.K. (1994) *Case Study Research*, 2nd edn. Beverly Hills, California: Sage.

Yoshida, R.K. (1980) Multidisciplinary decision making in special education: A review of the issue. Socia, *Psychology Review*, **9** (3), Summer, 221–227.

Yoshida, R.K., Fentond, K. and Maxwell, J. (1978) Group decision making in the planning team process: Myth or reality? *Journal of School Psychology*, **16**, 237–244.

Ysseldyke, J.E., Algozzine, B. and Mitchell, J. (1982) Special education team decision-making: An analysis of current practice, *Personnel and Guidance Journal*, **60** (5), January, 308–313.

Yukl, G. (1998) *Leadership in Organizations*, 4th edn. State University of New York at Albany: Prentice-Hall International.

Zahrly, J. and Tosi, H. (1989) The differential effects of organizational induction process on early work role adjustment, *Journal of Organizational Behavior*, **10**, 59–74.

Zakeri, M., Olomolaiye, P.O., Holt, G. and Harries, F.C. (1996) A survey of constraints on construction operative motivation, *Construction Management and Economics*, **14**, 417–426.

# Index